Robert Schaefer

Foundations of Global Genetic Optimization

Studies in Computational Intelligence, Volume 74

Editor-in-chief
Prof. Janusz Kacprzyk
Systems Research Institute
Polish Academy of Sciences
ul. Newelska 6
01-447 Warsaw
Poland
E-mail: kacprzyk@ibspan.waw.pl

Further volumes of this series
can be found on our homepage:
springer.com

Vol. 54. Fernando G. Lobo, Cláudio F. Lima
and Zbigniew Michalewicz (Eds.)
Parameter Setting in Evolutionary Algorithms, 2007
ISBN 978-3-540-69431-1

Vol. 55. Xianyi Zeng, Yi Li, Da Ruan and Ludovic Koehl
(Eds.)
Computational Textile, 2007
ISBN 978-3-540-70656-4

Vol. 56. Akira Namatame, Satoshi Kurihara and
Hideyuki Nakashima (Eds.)
Emergent Intelligence of Networked Agents, 2007
ISBN 978-3-540-71073-8

Vol. 57. Nadia Nedjah, Ajith Abraham and Luiza de
Macedo Mourella (Eds.)
*Computational Intelligence in Information Assurance
and Security,* 2007
ISBN 978-3-540-71077-6

Vol. 58. Jeng-Shyang Pan, Hsiang-Cheh Huang, Lakhmi
C. Jain and Wai-Chi Fang (Eds.)
Intelligent Multimedia Data Hiding, 2007
ISBN 978-3-540-71168-1

Vol. 59. Andrzej P. Wierzbicki and Yoshiteru
Nakamori (Eds.)
Creative Environments, 2007
ISBN 978-3-540-71466-8

Vol. 60. Vladimir G. Ivancevic and Tijana T. Ivacevic
*Computational Mind: A Complex Dynamics
Perspective,* 2007
ISBN 978-3-540-71465-1

Vol. 61. Jacques Teller, John R. Lee and Catherine
Roussey (Eds.)
Ontologies for Urban Development, 2007
ISBN 978-3-540-71975-5

Vol. 62. Lakhmi C. Jain, Raymond A. Tedman
and Debra K. Tedman (Eds.)
*Evolution of Teaching and Learning Paradigms
in Intelligent Environment,* 2007
ISBN 978-3-540-71973-1

Vol. 63. Wlodzislaw Duch and Jacek Mańdziuk (Eds.)
Challenges for Computational Intelligence, 2007
ISBN 978-3-540-71983-0

Vol. 64. Lorenzo Magnani and Ping Li (Eds.)
*Model-Based Reasoning in Science, Technology, and
Medicine,* 2007
ISBN 978-3-540-71985-4

Vol. 65. S. Vaidya, L.C. Jain and H. Yoshida (Eds.)
*Advanced Computational Intelligence Paradigms in
Healthcare-2,* 2007
ISBN 978-3-540-72374-5

Vol. 66. Lakhmi C. Jain, Vasile Palade and Dipti
Srinivasan (Eds.)
*Advances in Evolutionary Computing for System
Design,* 2007
ISBN 978-3-540-72376-9

Vol. 67. Vassilis G. Kaburlasos and Gerhard X. Ritter
(Eds.)
*Computational Intelligence Based on Lattice
Theory,* 2007
ISBN 978-3-540-72686-9

Vol. 68. Cipriano Galindo, Juan-Antonio
Fernández-Madrigal and Javier Gonzalez
*A Multi-Hierarchical Symbolic Model
of the Environment for Improving Mobile Robot
Operation,* 2007
ISBN 978-3-540-72688-3

Vol. 69. Falko Dressler and Iacopo Carreras (Eds.)
*Advances in Biologically Inspired Information Systems:
Models, Methods, and Tools,* 2007
ISBN 978-3-540-72692-0

Vol. 70. Javaan Singh Chahl, Lakhmi C. Jain, Akiko
Mizutani and Mika Sato-Ilic (Eds.)
Innovations in Intelligent Machines-1, 2007
ISBN 978-3-540-72695-1

Vol. 71. Norio Baba, Lakhmi C. Jain and Hisashi Handa
(Eds.)
*Advanced Intelligent Paradigms in Computer
Games,* 2007
ISBN 978-3-540-72704-0

Vol. 72. Raymond S.T. Lee and Vincenzo Loia (Eds.)
Computation Intelligence for Agent-based Systems, 2007
ISBN 978-3-540-73175-7

Vol. 73. Petra Perner (Ed.)
Case-Based Reasoning on Images and Signals, 2007
ISBN 978-3-540-73178-8

Vol. 74. Robert Schaefer
Foundations of Global Genetic Optimization, 2007
ISBN 978-3-540-73191-7

Robert Schaefer

Foundations of Global Genetic Optimization

Chapter 6 written by Henryk Telega

With 44 Figures

Prof. Robert Schaefer
Department of Computer Science
AGH University of Science and Technology
Mickiewicza 30
30-059 Kraków
Poland
E-mail: schaefer@agh.edu.pl

Chapter 6 written by:

Henryk Telega
Institute of Computer Science
Jagiellonian University
Nawojki 11
30-072 Kraków
Poland
E-mail: telega@softlab.ii.uj.edu.pl

Library of Congress Control Number: 2007929548

ISSN print edition: 1860-949X
ISSN electronic edition: 1860-9503
ISBN 978-3-540-73191-7 Springer Berlin Heidelberg New York

Springer is a part of Springer Science+Business Media
springer.com

Cover design: deblik, Berlin
Typesetting by the SPi using a Springer LaTeX macro package
Printed on acid-free paper SPIN: 11733416 89/SPi 5 4 3 2 1 0

Contents

List of Figures

List of Algorithms

1

Introduction

Genetic algorithms today constitute a family of effective global optimization methods used to solve difficult real-life problems which arise in science and technology. Despite their computational complexity, they have the ability to explore huge data sets and allow us to study exceptionally problematic cases in which the objective functions are irregular and multimodal, and where information about the extrema location is unobtainable in other ways.

They belong to the class of iterative stochastic optimization strategies that, during each step, produce and evaluate the set of admissible points from the search domain, called the random sample or population. As opposed to the Monte Carlo strategies, in which the population is sampled according to the uniform probability distribution over the search domain, genetic algorithms modify the probability distribution at each step.

Mechanisms which adopt sampling probability distribution are transposed from biology. They are based mainly on genetic code mutation and crossover, as well as on selection among living individuals. Such mechanisms have been tested by solving multimodal problems in nature, which is confirmed in particular by the many species of animals and plants that are well fitted to different ecological niches. They direct the search process, making it more effective than a completely random one (search with a uniform sampling distribution). Moreover, well-tuned genetic-based operations do not decrease the exploration ability of the whole admissible set, which is vital in the global optimization process.

The features described above allow us to regard genetic algorithms as a new class of artificial intelligence methods which introduce heuristics, well tested in other fields, to the classical scheme of stochastic global search.

Right from the beginning, genetic algorithms have aroused the interest of users and scientists, who try to apply them when solving engineering problems (optimal design, stuff deposit investigation and defectoscopy etc.) as well as to explain the essence of their complex behavior.

At least ten large international conferences are devoted to the theory and applications of genetic optimization. The most important and well established

R. Schaefer: *Foundation of Global Genetic Optimization*, Studies in Computational Intelligence (SCI) **74**, 1–6 (2007)
www.springerlink.com

seems to be FOGA[1] (Foundation of Genetic Algorithms), GECCO[2] (Genetic and Evolutionary Computation), PPSN[3] (Parallel Problem Solving from Nature) and CEC[4] (IEEE Congress on Evolutionary Computation). Almost all conferences in artificial intelligence, optimization, distributed processing, CAD/CAE and various branches of technology contain sessions or workshops that gather contributions showing specialized applications of genetic computing and their theoretical motivations. Many important events were organized by national associations, in particular the National Conference in Genetic Algorithms and Global Optimization KAEiOG[5] has taken place in Poland annually since 1996.

There are also several scientific journals devoted solely to genetic algorithm theory and applications. It is worth highlighting *IEEE Transactions on Evolutionary Computation*[6] and *Evolutionary Computation*[7] among them.

Besides the many research papers cited in this book, we would like to draw the reader's attention to monographs that try to comprehend results from various branches of genetic computation and trace new directions in research and applications. The pioneering book in this area, entitled *Adaptation in Natural and Artificial Systems*, was written by Holland [85] in 1975. The author defined binary genetic operations and related them to the real modifications and inheritance of genetic code. He also tried to deliver a formal description and quantitative evaluation of the artificial genetic process by formulating a popular schemata theorem. Important bibliographical items that show the number and variety of evolutionary optimization techniques, as well as more formal descriptions of algorithms, are the books of Goldberg 1989, [74], Michalewicz 1992, [110] and Koza 1992, [99]. Due to its intentions and large scope, the monograph of Bäck, Fogel and Michalewicz from 1997, [15] as well as its compressed and improved version [10, 11], is impressive. An exceptional book which discusses parallel models and implementations of genetic computation was written by Cantú-Paz 2000, [45]. Another title of this type, published by Osyczka 2002, [124] summarizes the genetic algorithm applications to multi-criteria design and optimization.

One well-known book, written in 1999 by Vose, [193] delivers the most important results concerning formal analysis of the so-called *Simple Genetic Algorithm* (SGA). His approach is based on SGA modeling as the Markov chain, whose trajectories are located in the space of states common for the class of algorithms with different population cardinality. The main results characterize the SGA asymptotic behavior by the number of iterations (genetic epochs) tending to infinity, as well as for an infinitely growing population size.

---

[1] http://www.sigevo.org/foga-2007/
[2] http://www.sigevo.org/gecco-2006/
[3] http://ls11-www.cs.uni-dortmund.de/PPSN/
[4] http://www.cec2007.org/
[5] http://kaeiog.elka.pw.edu.pl/
[6] http://ieee-cis.org/pubs/tec/
[7] http://www.mitpressjournals.org/loi/evco?cookieSet=1

The next important contributions in this area are the chapters written by Rudolph [143, 144, 145], Rudolph and Beyer [27], which are integral parts of three books, edited by Bäck, Fogel and Michalewicz [15, 10, 11]. They analyze the particular type of convergence of genetic algorithms with real number, phenotypic encoding.

There are many distinguished books, which have been printed recently, dealing with genetic algorithm theory. Spears 2000, [178] discusses the role of mutation and recombination, in algorithms with the binary genetic universum, in detail. Beyer 2001, [26] presents an exhaustive analysis of the evolution strategy progress using strong regularity assumptions with respect to fitness. Langdon and Poli 2002, [102] extend some theoretical results which come from binary schemata theory to the case of genetic programming with the genetic universum as a space of graphs. Reeves and Rowe 2003, [134] present a critical view of the various approaches for studying genetic algorithm theory and discuss the perspectives in this area.

Finally it worth mentioning two Polish books. The first one, written by Arabas [5], delivers the author's original approach and comments to selected genetic techniques, preceded by a broad mathematical description concerning single- and multi-criteria optimization problems. The second one [149] marks the beginnings of this book.

This work delivers a new approach for studying genetic algorithms by modeling them as dynamic systems which transform the probabilistic sampling measures (probability distributions on the admissible set of solutions) in a regular way. This approach allows us to show that genetic algorithms may effectively find the subsets of the search domain rather than isolated points, e.g. the central parts of the basins of attraction of the local minima rather than isolated minimizers. This feature reflects the character of elementary evolutionary mechanisms implemented here that lead to the whole flock (population) surviving by the fast exploration of new regions of feeding rather than care of the single flock member (individual). The attention of the reader will be turned to such kinds of genetic algorithm instances (e.g. two-phase methods, genetic sample clustering, and sensitivity analysis) for which the above features may guarantee that all solutions are found and the stopping rule is verified. It will also be shown that the traditional use of a genetic algorithm to solve local optimization problems may meet obstacles which arise from the inherent features of this group of methods.

We will focus on the ability of genetic algorithms to solve global, continuous optimization problems so that the admissible solutions make the regular subset (with the Lipschitz boundary) of the positive Lesbegue measure in a dense, finite dimensional linear-metric space. We do not consider genetic algorithm instances which can only solve discrete optimization problems. We will also omit such features of the common algorithm instances that are valid only for the discrete search domain.

Detailed definitions of standard continuous global optimization problems are given at the beginning of this book. Problems which involve finding all global extremes as well as the predefined class of local extremes are discussed in Section 2.1. New optimization problems, leading to recognizing and approximating sets which are the central parts of the basins of attraction of local minimizers, are also introduced and discussed in this section. In Sections 2.2, 2.3, a general, abstract scheme of the stochastic, population search and its basic qualitative, asymptotic features are specified. In the light of these formulations, the basic mechanisms of genetic computation, presented in Chapter 3, exhibit their real nature and working directions. Such an approach is perhaps much less mysterious than the traditional one based on biological analogy. It also allows the synthetic presentation of many details common for quite different algorithm classes (e.g. genetic algorithms with the finite set of codes and evolutionary strategies using phenotypic, real number encoding).

All standard genetic operations presented in Chapter 3 lead to the stationary rule of sampling adaptation i.e. the rule does not depend on the genetic epoch in which it is applied. In other words, the probability distribution utilized for sampling is obtained in the same way, taking only the current population into account. The class of genetic algorithms that permit only stationary adaptation rules will be called *self-adaptive genetic algorithms*, because the transition rules of sampling probability distribution are not modified by any external control or by any feedback signals coming from the previous population monitoring. The taxonomy and a short description of *adaptive genetic algorithms* that break the principle of stationary sampling probability transition are given in Chapter 5.

The core of this book, located in Chapter 4, synthesizes the mathematical models of genetic algorithm dynamics and their asymptotic features. The main results presented in this chapter are based on the stochastic model that operates on the space of states whose elements are populations or their unambiguous representations. Genetic algorithms are assumed to be self-adaptive, which implies the uniform Markovian rule of state transition.

Asymptotic results obtained for the Simple Genetic Algorithm (see Section 4.1.2) are based on features of the so-called *genetic operator*, introduced by Vose and his co-workers, sometimes called the SGA heuristics (see e.g. [193]). In the same section, theorems concerning the transformation of sampling measures, and their transport from the space of states of the genetic algorithm to the search domain, are considered. These results motivate the application of genetic algorithms in searching and approximating the central parts of the basins of attractions of the local, isolated minimizers. Such results are also helpful in the analysis of two-phase global optimization strategies which utilize genetic algorithms during the first, exploration phase (see Chapter 6). Section 4.1.3 contains results of the Markov theory of evolutionary algorithms $(\mu + \lambda)$-type with the elitist selection.

Finally, Section 4.2 comprehends asymptotic results obtained for genetic algorithms with very small populations and Section 4.3 delivers some comments which lead to the precise formulation and verification of the schemata theorem for SGA.

An important part of this book, written by Henryk Telega, is located in Chapter 6. It delivers a survey of two-phase stochastic global optimization methods. Such strategies consist of finding the approximation of extreme attractors in the first phase, called *global phase*, and pass to the detailed, local search in each attractor in the second phase, called *local phase*. Probabilistic asymptotic correctness as well as stopping rules of two-phase strategies are also discussed.

A new global phase strategy, called *Clustered Genetic Search* (CGS), which utilizes the genetic sample to recognize the central parts of attractors, is introduced. The advantageous features of genetic algorithms that regularly transform sampling measures to ones which become denser close to the local extrema guarantee the proper definition of such a strategy. In particular, by using theorems formulated in Chapter 4, the probabilistic asymptotic correctness and the stopping rule of CGS are verified.

We have omitted detailed, technical proofs of some cited theorems, remarks and formulas due to the necessary limitation of volume and the assumed engineering profile of this book. Readers are extensively referred to sources in each particular case.

Several important computation examples are placed in Section 2.4 in order to demonstrate the skill of genetic algorithms in solving optimal design problems, formulated as continuous global optimization ones. The second group of tests exhibits characteristic features of the clustered genetic search (CGS) running for a small set of classical multimodal benchmarks (see Chapter 6).

Readers only require a basic mathematical preparation and a maturity at a level typical for the MS courses in science, especially in the area of real valued function analysis, linear algebra as well as probability theory and stochastic processes.

The book may be recommended in particular to readers who have a basic insight into genetic-based computational algorithms and are looking for an explanation of their quantitative features and for their advanced applications and further development. It may be helpful to engineers in solving difficult global optimization problems in technology, economics and the natural sciences. This is especially true in the cases of multimodality, weak regularity of the objective function and large volumes of the search domain. It may also inspire researchers employed in studying stochastic optimization and artificial intelligence.

I would like to thank everyone who has played a part in the preparation of this book. Special thanks are directed to Professors Tadeusz Burczyński, Iwona Karcz-Dulęba and Katarzyna Adamska who have permitted me to enclose valuable results and computational examples, published in their research papers. I am grateful to Professors Krzysztof Malinowski, Roman Galar, Mariusz Flasiński, Zdzisław Denkowski, Marian Jabłoński, Jarosław Arabas and Kazimierz Grygiel for their critical view of ideas presented in the manuscript and their many useful and detailed suggestions.

Finally, I would like to thank my wife Irena for her constant support and considerable help.

Robert Schaefer

2

Global optimization problems

This chapter introduces readers to the world of continuous global optimization problems. We start with detailed definitions of the search space, admissible domain and objective function. The most conventional problem, which consists of finding all admissible points in which the objective function has its global extreme, is the basis of further considerations. The next problems concern finding local extremes. We also consider the approximate problems that allow finite accuracy of data representation. Much space is devoted to the definition of the basins of attraction of local isolated extremes and the problem of their approximate recognition. Next we introduce the scheme of population-oriented stochastic global optimization search. Two important instances: random walk and Pure Random Search are defined. We have focused on a more formal definition of populations (random samples) and the mathematical operations on them. Moreover, definitions that classify search possibilities and some kind of convergence are formulated and commented on. The chapter contains several computation examples, which show the potential skill of various genetic global optimization strategies by solving difficult continuous engineering problems.

2.1 Definitions of global optimization problems

Let us denote by V the *space of solutions* which is the complete metric space with a distance function $d : V \times V \to \mathbb{R}_+$ and a topology $\text{top}(V)$ induced by this metric (see e.g. [162]). Here $\mathbb{R}_+ = \{x \in \mathbb{R}; x \geq 0\}$ and $\text{top}(V)$ is the family of open sets in V. The space V is a dense set i.e. each point $x \in V$ is the concentration point in V.

We also impose that V is the space of points of a finite dimensional affine structure. More precisely, the "space of directions" $\hat{V}$ exists which is the finite dimensional Hilbert space ($\dim(\hat{V}) = N < +\infty$) and two mappings so that:

- the first of them, $V \times V \ni (x, y) \to y - x \in \hat{V}$ assigns the "joining" vector to the ordered pair of points,

R. Schaefer: *Foundation of Global Genetic Optimization*, Studies in Computational Intelligence (SCI) **74**, 7–30 (2007)
www.springerlink.com

- the second one, $V \times \hat{V} \ni (x, v) \to x + v \in V$ translates the point by the vector.

We assume moreover, that the distance function induced by the Euclidean norm $\|\cdot\|_2$ in $\hat{V}$ is topologically equivalent to the original one in V

$$\begin{gathered} \exists\, \alpha_1, \alpha_2 > 0;\ \forall x, y \in V\ \alpha_1 \|y - x\|_2 \le d(y, x) \le \alpha_2 \|y - x\|_2 \\ \|v\|_2 = \sqrt{(v, v)},\ v \in \hat{V} \end{gathered} \tag{2.1}$$

where $(\cdot, \cdot)$ stands for the scalar product in $\hat{V}$. A detailed definition of the affine structure may be found in many books in linear algebra and geometry, particularly in Spivak [179].

Such an extended structure of the search space provides us with all the necessary tools for analyzing optimization problems and methods. In particular, we are able to consider the neighborhood of the extreme $x^* \in V$ as a ball with respect to the metric d, as well as study the convergence of the minimization sequence $\{x_i\} \subset V$ with respect to the topology $\mathrm{top}(V)$. Moreover, we may precisely define the meaning of searching in the direction $v \in \hat{V}$ starting from $x_0 \in V$ which results in the new point $x_0 + v \in V$, while the step length is $\|v\|_2$.

In almost all the cases discussed later the space of solutions and the direction space will be supported by the same set $\mathbb{R}^N$ (both $V, \hat{V}$ are $\mathbb{R}^N$), so we will refer all elements of the affine structure to V for the sake of simplicity. Detailed comments will be delivered in other cases.

The subset $\mathcal{D} \subset V$ will denote the *set of admissible solutions* of each optimization problem defined later. We will frequently assume that $\mathcal{D}$ is compact, which implies that $\mathcal{D}$ is bounded ($\mathrm{diam}(\mathcal{D}) < +\infty$) and closed in the finite dimensional case. In several cases the Lipschitz boundary of the admissible set will be required, i.e. $\partial\mathcal{D}$ will be a finite composition of $(N-1)$-dimensional, C^1-regular hypersurface pieces, continuously glued, without infinitely narrow edges (blades) (see e.g. Zeidler [207] for details).

Definition 2.1. *The* objective function *is the well defined, bounded mapping*

$$\Phi : \mathcal{D} \to \mathbb{R}_+;\ 0 \le \Phi(x) \le M < +\infty,\ \forall x \in \mathcal{D} \tag{2.2}$$

that evaluates admissible points from $\mathcal{D}$. □

The well defined function on $\mathcal{D}$ means the function which is countable for every $x \in \mathcal{D}$ (see e.g. Cromen, Leiserson, Rivest [51], Manna [108]).

In some situations the classical, continuous differentiability up to the second order of the objective function will be necessary, which will be denoted as $\Phi \in C^l(\mathcal{D})$, $l = 0, 1, 2$. Because $\mathcal{D}$ may be closed, then we assume $\Phi \in C^l(A)$, $l = 0, 1, 2$ for some open set $A \in \mathrm{top}(V)$ and $\mathcal{D} \subset A$ in such cases.

We will consider four basic global optimization problems:

Problem $\mathbf{\Pi_1}$: *Find all points $x^* \in \mathcal{D}$ so that*

$$\Phi(x) \leq \Phi(x^*), \; \forall x \in \mathcal{D} \tag{2.3}$$

being global maximizers for the objective function Φ on $\mathcal{D}$. □

Problem $\mathbf{\Pi_2}$: *Find all points $x^+ \in \mathcal{D}$ so that*

$$\exists A \in \text{top}(V); \; x^+ \in A, \; \Phi(x) \leq \Phi(x^+), \; \forall x \in A \cap \mathcal{D} \tag{2.4}$$

being local maximizers for the objective function Φ on $\mathcal{D}$. □

Problem $\mathbf{\Pi_3}$: *Find all points $x^+ \in \mathcal{D}$ so that*

$$\exists A \in \text{top}(V); \; x^+ \in A, \; \Phi(x) < \Phi(x^+), \; \forall x \in A \cap \mathcal{D} \tag{2.5}$$

being local isolated maximizers for the objective function Φ on $\mathcal{D}$. □

Taking into consideration the restricted accuracy of real computations, the following alternative, approximate global optimization problems may be defined together:

Problem $\mathbf{\Pi_1^{a1}}(\mathbf{\Pi_2^{a1}}, \mathbf{\Pi_3^{a1}})$: *For each point x^* being a solution to $\mathbf{\Pi_1}$ (for each point x^+ being a solution to $\mathbf{\Pi_2}, \mathbf{\Pi_3}$) find at least one point $x \in A_\varepsilon(x^*)$* ($x \in A_\varepsilon(x^+)$), where

$$A_\varepsilon(y) = \{x \in V; \|x - y\| < \varepsilon\} \cap \mathcal{D} \tag{2.6}$$

for an arbitrary $\varepsilon > 0$. □

The main disadvantage of the above definition is its dependency on the norm $\|\cdot\|$ in the space V. We may alternatively define

$$A_\varepsilon(y) = \{x \in \mathcal{D}; \Phi(x) \geq \Phi(y) - \varepsilon\} \tag{2.7}$$

without using a norm in V. The problems obtained in this way, denoted as $\mathbf{\Pi_1^{a2}}, \mathbf{\Pi_2^{a2}}, \mathbf{\Pi_3^{a2}}$, have quite a different structure than the previous ones ($\{\mathbf{\Pi_i^{a1}}\}, i = 1, 2, 3$).

Following the idea presented by Rinnooy Kan, Timmer [137] and Betró [22] the another definition for the extrema neighboring sets may be suggested

$$\begin{aligned} A_\varepsilon(y) &= \{x \in \mathcal{D}; \phi(\Phi(x)) \leq \varepsilon\}, \\ \phi(z) &= \frac{\text{meas}(\{\xi \in \mathcal{D}; \Phi(\xi) \geq z\})}{\text{meas}(\mathcal{D})} \end{aligned} \tag{2.8}$$

where meas denotes the Lesbegue measure on V. In this way we obtained the new group of approximate problems denoted by $\mathbf{\Pi_1^{a3}}, \mathbf{\Pi_2^{a3}}, \mathbf{\Pi_3^{a3}}$. The last two propositions $\{\mathbf{\Pi}_i^{\mathbf{a}j}\}, i = 1,2,3; j = 2,3$ may exhibit substantial inaccuracy, because for $y = x^*$ (or $y = x^+$) the set $A_\varepsilon(y)$ defined by formula 2.7 (or 2.8) may be not a connected set.

All the global optimization problems $\{\mathbf{\Pi}_i, \mathbf{\Pi}_i^{\mathbf{a}j}\}, i = 1,2,3; j = 1,2,3$ already defined lead to finding global or local maximizers or their sufficiently closed neighborhood. One can also define the analogous problems $\{\tilde{\mathbf{\Pi}}_i, \tilde{\mathbf{\Pi}}_i^{\mathbf{a}j}\}, i = 1,2,3; j = 1,2,3$ that consist of finding global or local minimizers for the objective function Φ on the set $\mathcal{D}$. The meaning of symbols $x^*, x^+, A_\varepsilon, \phi$ will depend on the current context.

Remark 2.2. Let Φ be the objective function associated with one arbitrary global maximization problem from the class $\{\mathbf{\Pi}_i, \mathbf{\Pi}_i^{\mathbf{a}j}\}, i = 1,2,3; j = 1,2,3$ discussed above. We may establish the equivalent global minimization problem from the class $\{\tilde{\mathbf{\Pi}}_i, \tilde{\mathbf{\Pi}}_i^{\mathbf{a}j}\}, i = 1,2,3; j = 1,2,3$ by setting a new objective function $\tilde{\Phi} = M - \Phi$ where M is the upper bound for Φ on the set $\mathcal{D}$. Moreover the formulas 2.7, 2.8 have to be rewritten to the form

$$A_\varepsilon(y) = \left\{x \in \mathcal{D}; \tilde{\Phi}(x) \leq \tilde{\Phi}(y) + \varepsilon\right\} \tag{2.9}$$

$$\begin{aligned} A_\varepsilon(y) &= \left\{x \in \mathcal{D}; \phi(\tilde{\Phi}(x)) \leq \varepsilon\right\}, \\ \phi(z) &= \frac{\text{meas}(\{\xi \in \mathcal{D}; \tilde{\Phi}(\xi) \leq z\})}{\text{meas}(\mathcal{D})} \end{aligned} \tag{2.10}$$

originally introduced by Rinnooy Kan, Timmer [137] and Betró [22]. The formula 2.6 is also valid for the minimization problems.

Similarly, for $\tilde{\Phi}$, the objective function of the minimization problem that satisfies Definition 2.1, the analogous maximization problem may be obtained by setting $\Phi = M - \tilde{\Phi}$, where M stands for the upper bound for $\tilde{\Phi}$ in this case.

It should be underlined that the constant M must be well known in both cases. □

Of course, the way of obtaining equivalent minimization and maximization problems presented above is not a unique one, but we will restrict ourselves to the problems which satisfy the assumptions of Remark 2.2 for the sake of simplicity.

Remark 2.3. All considerations contained in Section 2.1 may be extended to the class of well defined functions $\Psi : \mathcal{D} \to \mathbb{R}$ bilaterally bounded $m \leq \Psi(x) \leq M$, $\forall x \in \mathcal{D}$ for some finite $m, M \in \mathbb{R}$. It is enough to set $\Phi = \Psi + \max\{0, -m\}$ in order to satisfy the conditions of definition 2.1. Similarly, as in Remark 2.2 we have to know the values of both constants m, M the lower and upper bounds of Ψ on the set $\mathcal{D}$. □

In the remaining part of this section we will study minimization problems for which the equivalent maximization ones may be established. All problems from the class $\{\tilde{\mathbf{\Pi}}_i, \tilde{\mathbf{\Pi}}_i^{\mathbf{a}j}\}, i = 1,2,3; j = 1,2,3$ which satisfy the assumptions of Remarks 2.2, 2.3 are good candidates. The function to be minimized that constitutes the objective of these problems will be denoted by $\tilde{\Phi} : \mathcal{D} \to \mathbb{R}_+$.

Further considerations need to assume that the proper local optimization method loc exists. It can be started from an arbitrary point $x_0 \in \mathcal{D}$ and then generates the sequence of points in $\mathcal{D}$ which always converges to some $\mathrm{loc}(x_0) \in \mathcal{D}$, which is the local minimizer attainable from the starting point x_0. The local method may be interpreted as the mapping $\mathrm{loc} : \mathcal{D} \ni x_0 \to \mathrm{loc}(x_0) \in \mathcal{D}$.

Next we will distinguish the important group of local methods (see Rinnooy, Kan, Timmer [137], Dixon Gomulka Szegö [61]).

Definition 2.4. *The local method* loc *will be called strictly descent on $\mathcal{D}$ if for each starting point $x_0 \in \mathcal{D}$ and an arbitrary norm $\|\cdot\|$ in V it generates the sequence $\{x_i\}, i = 0,1,2\ldots \subset \mathcal{D}$ so that*

$$x_{i+1} = x_i + \alpha_i p_i \;\forall i = 0,1,2\ldots \;\|p_i\| = 1,\; \alpha_i \geq 0 \tag{2.11}$$

Moreover, the sequence converges to $x^+ = \mathrm{loc}(x_0)$, the local minimizer of the objective function $\tilde{\Phi}$, and satisfies

$$\forall i = 0,1,2\ldots \;\forall \alpha,\beta;\; 0 \leq \alpha < \beta \leq \alpha_i \Rightarrow \tilde{\Phi}(x_i + \beta p_i) \leq \tilde{\Phi}(x_i + \alpha p_i) \tag{2.12}$$

□

Definition 2.5. *Let* loc *be the strictly descent method on $\mathcal{D}$ and x^+ the isolated local minimizer of $\tilde{\Phi}$ in $\mathcal{D}$. The set*

$$R_{x^+}^{loc} = \left\{x \in \mathcal{D}; \mathrm{loc}(x) = x^+\right\} \tag{2.13}$$

will be called the set of attraction of x^+ with respect to the local method loc. *We will frequently simplify its notation to R_{x^+}.* □

The above definition follows the idea given by Zieliński [208]. Next we introduce four important quantities:

- $L(y) = \{x \in \mathcal{D}; \tilde{\Phi}(x) \leq y\}$ the *level set* of the objective function $\tilde{\Phi}$.
- $L_x(y)$ the connected part of $L(y)$ that contains x.
- $$\bar{y}_{x^+} = \begin{cases} \inf \left\{ \begin{array}{c} y : \exists z^+ \text{ local minimizer of } \tilde{\Phi}, \\ z^+ \neq x^+, z^+ \in L_{x^+}(y) \end{array} \right\} & \text{if } z^+ \text{ exists} \\ \max_{x \in \mathcal{D}}\{\tilde{\Phi}(x)\} & \text{in the other case} \end{cases}$$
  the *cutting level* associated with an arbitrary, isolated local minimizer x^+. This description will be further simplified to $\bar{y}$ if the context is clear.
- $\tilde{L}_{x^+}(\bar{y}) = \left\{x \in L_{x^+}(\bar{y}); \tilde{\Phi}(x) < \bar{y}\right\}$

Definition 2.6. *The basin of attraction $\mathcal{B}_{x^+} \subset \mathcal{D}$ of a local isolated minimizer x^+ of the function $\tilde{\Phi} : \mathcal{D} \to \mathbb{R}_+$ is the connected part of $\tilde{L}_{x^+}(\bar{y})$ that contains x^+. The basin of attraction $\mathcal{B}_{x^+} \subset \mathcal{D}$ of a local isolated maximizer x^+ of the function $\Phi : \mathcal{D} \to [0, M] \subset \mathbb{R}_+$; $M < +\infty$ is the basin of attraction of the local isolated minimizer x^+ of the function $\tilde{\Phi} = M - \Phi$.* □

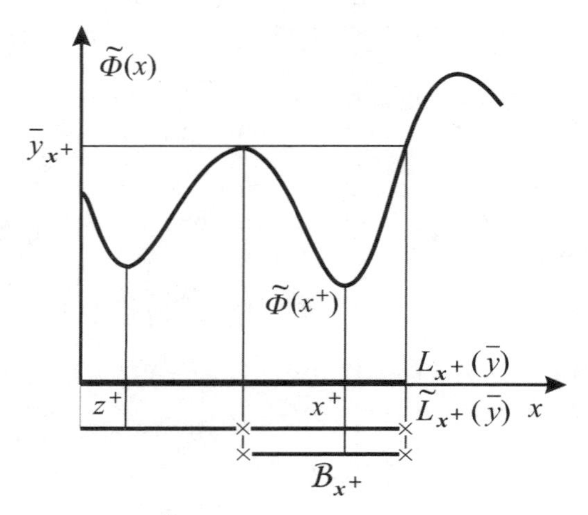

Fig. 2.1. Level sets $L_{x^+}(\bar{y}), \tilde{L}_{x^+}(\bar{y})$ and the basin of attraction $\mathcal{B}_{x^+}$ of the local isolated minimizer x^+ for the function $\tilde{\Phi} : \mathbb{R} \to \mathbb{R}_+$.

Remark 2.7. (see Rinnooy Kan, Timmer [137], Dixon, Gomulka, Szegö [61]) Every strictly descent method loc has the following features:

1. Let $x \in \mathcal{D}$ be an arbitrary admissible point and $y \leq \tilde{\Phi}(x)$ then $\text{loc}(x_0) \in L_x(y)$ for all starting points $x_0 \in L_x(y)$.
2. $\text{loc}(x_0) = x^+$ for all starting points $x_0 \in \mathcal{B}_{x^+}$.
3. $\mathcal{B}_{x^+} \subset R^{loc}_{x^+}$. □

Figure 2.1 shows the differences between $L(\bar{y}), \tilde{L}_{x^+}(\bar{y})$ and $\mathcal{B}_{x^+}$ in the one dimensional case $\mathcal{D} \subset \mathbb{R}$.

We are ready now to define one more global optimization problem that consists of finding the approximation to the central parts of the basins of attraction.

Problem $\mathbf{\Pi_4}$: *For each isolated local maximizer $x^+ \in \mathcal{D}$ so that* $\text{meas}(\mathcal{B}_{x^+}) > 0$, *being a solution of problem* $\mathbf{\Pi_3}$ *find the closed set $C \subset \mathcal{B}_{x^+}$ so that* $\text{meas}(C) > 0$, and $x^+ \in C$. □

Similarly, as in Remark 2.2, the problem $\tilde{\mathbf{\Pi}}_\mathbf{4}$ associated with $\mathbf{\Pi_4}$ that consists of finding the approximation to the basins of attraction of local minimizers may be defined.

The importance of the problems defined above may be motivated twofold:

1. Stochastic, population based global optimization methods are well suited for solving such kind of problems. This hypothesis may be roughly motivated by the analysis of sampling measure dynamics. Sampling measures defined on the admissible set $\mathcal{D}$ tend toward the ones that become more dense close to the local minimizers. Such behavior is observed for some genetic algorithms and for the Monte Carlo algorithms equipped with some additional mechanisms. The detailed motivation will be delivered in Chapters 4 and 6.
2. The solution of problems $\mathbf{\Pi_4}$, $\tilde{\mathbf{\Pi}}_\mathbf{4}$ allow for effective solution of problems $\{\tilde{\mathbf{\Pi}}_i, \tilde{\mathbf{\Pi}}_i^{\mathbf{a}j}\}, i = 1,2,3; j = 1,2,3$ if the objective function is sufficiently regular (usually the C^2 regularity is required) in the neighborhood of the local minimizers. After the central parts of the basins of attraction have been recognized, the single, accurate local optimization method may be run in each basin. Local methods may be process in parallel. This approach will be discussed in Chapter 6.

2.2 General schema of a stochastic search

If optimization problems $\mathbf{\Pi_1}$ or $\tilde{\mathbf{\Pi}}_\mathbf{1}$ additionally satisfy the conditions that guarantee the existence and uniques of solution, as well as conditions that imply the convergence, and verify the stopping rule of local method (usually based on the evaluation of the norms of gradient and the Hesse matrix of the objective function Φ, see e.g. [66]), there is no need to use the stochastic methods, such as genetic algorithms, that are characterized by a far worse ratio between accuracy and computational cost than the local ones.

Real life optimization problems are usually much less regular. In particular they may exhibit the following irregularities:

1. The objective function Φ in not globally convex. More than one point in which the objective function has its local or global extremes exists in the admissible domain $\mathcal{D}$.
2. The objective function Φ is not globally continuously differentiable up to the second order ($\Phi \notin C^2(\mathcal{D})$). Sometimes, it preserves the regularity $C^i(A)$, $i = 0,1,2$ for A as the neighborhood of some local extremes.
3. The gradient vector and the Hesse matrix are difficult or impossible to compute explicitly. Complex approximation procedures have to be applied.

4. We do not have enough information about the "geography" of extreme locations in the admissible domain $\mathcal{D}$ in order to run more local methods, approximately one for each extreme (the multistart approach).

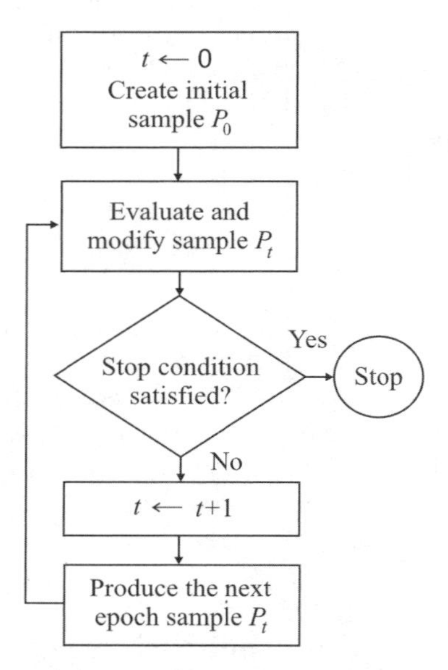

Fig. 2.2. General schema of the population-oriented, stochastic global optimization search.

The situations described above usually make stochastic, population oriented strategies much more efficient than the local, convex methods. The "skeleton" of such a strategy is shown in Figure 2.2.

We will denote by $\{P_t\}, t = 0, 1, 2, \ldots$ the consecutive random samples called *populations*. Populations are the multisets of "clones" from the admissible domain $\mathcal{D}$, which roughly means that a single point $x \in \mathcal{D}$ may have more than one "clone" in the particular population.

There are several ways to formalize the definition of multisets.

Definition 2.8. (see e.g. [168]) *Let $\mathcal{Z}$ be the set of patterns and $\eta : \mathcal{Z} \to \mathbb{Z}_+$ the occurrence function. The multiset A of elements (clones) from $\mathcal{Z}$ is the pair $A = (\mathcal{Z}, \eta)$ and may be understood as the set to which belong $\eta(x)$ clones of the pattern $x \in \mathcal{Z}$. The cardinality of the multiset is given by $\#A = \sum_{x \in \mathcal{Z}} \eta(x)$.*

□

The definition above allows us to introduce simple two-argument operations on multisets analogous to union, intersection, subtraction and the Cartesian product of sets.

Definition 2.9. *Let us consider two multisets $A_1 = (\mathcal{Z}, \eta_1)$ and $A_2 = (\mathcal{Z}, \eta_2)$.*

- *The union of multisets will be defined as* $A_1 \cup A_2 = (\mathcal{Z}, \eta)$ *so that* $\eta(x) = \eta_1(x) + \eta_2(x)\ \forall x \in \mathcal{Z}$.
- *The intersection of multisets is defined as* $A_1 \cap A_2 = (\mathcal{Z}, \eta)$ *so that* $\eta(x) = \min\{\eta_1(x), \eta_2(x)\}\ \forall x \in \mathcal{Z}$.
- *If* $\eta_1(x) \geq \eta_2(x)\ \forall x \in \mathcal{Z}$ *then the subtraction of* A_2 *from* A_1 *may be defined as* $A_1 \setminus A_2 = (\mathcal{Z}, \eta)$ *so that* $\eta(x) = \eta_1(x) - \eta_2(x)\ \forall x \in \mathcal{Z}$. *The opposite subtraction* $A_2 \setminus A_1$ *does not make sense in this case.*
- *The cartesian product of multisets will be defined as* $A_1 \times A_2 = (\mathcal{Z}^2, \eta)$ *so that* $\eta(x, y) = \eta_1(x)\eta_2(y)\ \forall (x, y) \in \mathcal{Z}^2$. □

Moreover, the inclusion relation may be extended to the multiset case.

Definition 2.10. *Let us consider two multisets* $A_1 = (\mathcal{Z}, \eta_1)$ *and* $A_2 = (\mathcal{Z}, \eta_2)$.

- *The inclusion of multisets* $A_1 \subseteq A_2$ *holds if* $\forall x \in \mathcal{Z}\ \eta_1(x) \leq \eta_2(x)$.
- *The strict inclusion of multisets* $A_1 \subset A_2$ *holds if* $\forall x \in \mathcal{Z}\ \eta_1(x) \leq \eta_2(x)$ *and* $\exists y \in \mathcal{Z}$ *so that* $\eta_1(x) < \eta_2(x)$. □

Remark 2.11. If the cardinality of the multiset $A = (\mathcal{Z}, \eta)$ is finite ($\#A < +\infty$) then the representation given by definition 2.8 is equivalent to the following $A = (\mathcal{Z}, \tilde{\eta})$ where $\tilde{\eta} : \mathcal{Z} \to [0, 1]$ turns back the occurrence frequency of the particular pattern, i.e. $\tilde{\eta}(x) = \eta(x)/\#A,\ \forall x \in \mathcal{Z}$. Obviously $\sum_{x \in \mathcal{Z}} \tilde{\eta}(x) = 1$. Please note that the assumption $\#\mathcal{Z} < +\infty$ is unnecessary in this case. □

Remark 2.12. The multiset representation $A = (\mathcal{Z}, \tilde{\eta})$ allows us to study multisets with infinite cardinality ($\#A = +\infty$) by setting

$$\tilde{\eta}(x) = \lim_{\#A \to +\infty} \frac{\eta(x)}{\#A}$$

It is unlikely such a representation is ambiguous because the arbitrary finite number of clones of $x \in \mathcal{Z}$ in A gives $\tilde{\eta}(x) = 0$ frequency in the limit case $\#A = +\infty$. □

Although the multiset representation given by definition 2.8 is well suited for multiset calculations, this approach has some disadvantages. In particular, it makes it impossible to distinguish two clones of the same pattern that belong to A. Sometimes we need to handle multisets like the regular sets in which two clones of the single pattern constitute separate elements. This concept presented below is based on the *permutational power* of the set used in topology introduced in the early papers of Smith [173] and Richardson [136]. More rigorous definitions and some interesting features of permutational power were delivered by Kwietniak [101].

Assuming the final cardinality $\#A = \mu < +\infty$ the alternative way of multiset definition (see e.g. [93, 149]) is based on the equivalence $eqp \subset (\mathcal{Z}^\mu)^2$ so that:

$$\begin{aligned} &\forall (z_1, \dots, z_\mu), (y_1, \dots, y_\mu) \in \mathcal{Z}^\mu \\ &(z_1, \dots, z_\mu)\, eqp\, (y_1, \dots, y_\mu) \Leftrightarrow \exists \sigma \in S_\mu; z_i = y_{\sigma(i)},\ i = 1, \dots, \mu \end{aligned} \tag{2.14}$$

where S_μ denotes the group of permutations of the μ-element set.

The simple motivation that *eqp* is really the equivalence relation is left as the reader's exercise.

Definition 2.13. *The multiset A of elements (clones) from $\mathcal{Z}$ of final cardinality $\mu < +\infty$ is a member of the quotient set $A \in \mathcal{Z}^\mu / eqp$.* □

The above definition 2.13 informs us that the ordering of elements in A is insignificant, and we should take the abstraction class of the μ-dimensional vector from $\mathcal{Z}^\mu$ with respect to *eqp* as the multiset model. This second multiset definition is rather difficult for multiset operations like those mentioned in definitions 2.9, 2.10, but will be helpful for introducing the useful multiset notation.

Remark 2.14. Both definitions 2.8, 2.13 are equivalent for finite multisets. □

Proof. We have to show that it is possible to introduce the one-to-one mapping that assigns each multiset A represented as the pair $(\mathcal{Z}, \eta)$ to the equivalence class from $\mathcal{Z}^\mu / eqp$. The occurrence function satisfies $\chi = \#\mathrm{supp}(\eta) \leq \mu < +\infty$, where $supp(\eta) = \{y \in \mathcal{Z}; \eta(y) > 0\}$. Then support η may be represented as $\{z_1, \dots, z_\chi\}$. Let us introduce

$$w = (w_1, \dots, w_\mu) \in \mathcal{Z}^\mu; w_i = \begin{cases} z_1,\ i = 1, \dots, \eta(z_1) \\ z_2,\ i = \eta(z_1) + 1, \dots, \eta(z_2) \\ \dots \ \dots \\ z_\chi,\ i = \eta(z_{\chi-1}) + 1, \dots, \eta(z_\chi) \end{cases} \tag{2.15}$$

Then we may assign $[w]_{eqp}$ to $(\mathcal{Z}, \eta)$, so the mapping $(\mathcal{Z}, \eta) \rightarrow [w]_{eqp}$ is well defined. We may easily check that this mapping is onto (is surjective). Let us take the arbitrary $\omega \in \mathcal{Z}^\mu / eqp$ and $w \in \omega$ for which we may uniquely define the function $\eta : \mathcal{Z} \ni x \rightarrow \#\{i; w_i = x\} \in \mathbb{Z}_+$. Such a definition is invariant with respect to the coordinate permutations of w, so η is also uniquely assigned to the whole ω. Now let us assume that there is $\omega \in \mathcal{Z}^\mu / eqp$ assigned to two pairs $(\mathcal{Z}, \eta_1)$, $(\mathcal{Z}, \eta_2)$. Immediately from the construction of ω (see 2.15) $\mathrm{supp}(\eta_1) = \mathrm{supp}(\eta_2) = \mathcal{Z}_A$ and for each $x \in \mathcal{Z}_A$, $\eta_1(x) = \eta_2(x)$, so $\eta_1 = \eta_2$ which proves the injectivity of the mapping. □

Remark 2.15. The membership relation has to be commented upon for both multiset representations given by the definitions 2.8, 2.13. If $x \in \mathcal{Z}$ will be the member of the multiset $A = (\mathcal{Z}, \eta)$, then of course $x \in \mathrm{supp}(\eta)$, which constitutes the necessary condition for multiset membership relation which can be drawn from the definition 2.8.

For multisets of finite cardinality $\#A = \mu < +\infty$ the definition 2.13 allows the representation $[w]_{eqp}$ of the multiset A, where w is given by formula 2.15.

The relation $x \in A$ means that x is one of the values of the finite sequence w which is the instance of $A \in \mathcal{Z}^\mu/eqp$. We may introduce the intuitive notation for the multiset A that will support many further considerations:

$$\left\langle \underbrace{z_1, z_1, \ldots, z_1}_{\eta(z_1)\text{ times}}, \underbrace{z_2, z_2, \ldots, z_2}_{\eta(z_2)\text{ times}}, \ldots, \underbrace{z_\chi, z_\chi, \ldots, z_\chi}_{\eta(z_\chi)\text{ times}} \right\rangle \tag{2.16}$$

where $\mathrm{supp}(\eta) = \{z_1, \ldots, z_\chi\}$, $\chi \leq \mu < +\infty$. Angle brackets $< >$ play a similar role in the multiset notation as curl brackets $\{\ \}$ in the set notation. They enclose the list of elements while the order of elements is insignificant. This notation allows us to distinguish two members of A that represent the same pattern, however its mathematical correctness is far from ideal. □

After the necessary, broad digression about multiset formalisms we turn back to stochastic strategies of global optimization.

If $\#P_t = 1, t = 0, 1, 2, \ldots$ (each sample is the singleton set) then we get perhaps the simplest strategy called *random walk*. This algorithm produces the maximization (minimization) sequence

$$x_0, x_1, x_2, \ldots \text{ such that } x_t \in \mathcal{D},\ \{x_t\} = P_t,\ t = 0, 1, 2, \ldots \tag{2.17}$$

the features of which completely differ from features that characterize sequences obtained by local convex optimization methods. In particular the sequence is not necessary monotone. Examples of random walk algorithms will be discussed in Section 4.2.1.

Let us comment now more extensively on the steps of stochastic strategy depicted in Figure 2.2.

Creation of the initial sample P_0 consists of sampling, usually using multiple sampling procedure, some points from the set $\mathcal{D}$. We will denote by $\mathcal{M}(\mathcal{D})$ the space of probabilistic measures on the admissible domain $\mathcal{D}$. The probability distribution $h_0 \in \mathcal{M}(\mathcal{D})$ is usually uniform on $\mathcal{D}$ or close to the uniform one.

$$h_0(A) = \frac{\mathrm{meas}(A)}{\mathrm{meas}(\mathcal{D})}, \text{ where } A \subset \mathcal{D} \text{ is the measurable set} \tag{2.18}$$

Such sampling reflects the typical situation appearing at the start of global optimization searches when no sampling regions of $\mathcal{D}$ are preferred because no information of extremes is available a'priori. When the information about the extreme locations is accessible at the start of computation, h_0 may be concentrated in some regions of $\mathcal{D}$.

In computational practice the initial sample P_0 may be obtained by using computer generators which deliver pseudo-random numbers with a uniform distribution on [0, 1]. Another possibility is to use the Halton, Sobol or Foure algorithms that generate the sequences of low discrepancy. More information concerning initial sample generation may be found in the papers of Arabas and Słomka [8] and Arabas [5].

Evaluation of the random sample P_t lies in computing effectively the values of the objective function for all sample members i.e. $\{\Phi(x)\}$, $x \in P_t$. If the sample is a multiset represented by $P_t = (\mathcal{D}, \eta_t)$ (see definition 2.8) then it is enough to compute $\{\Phi(x)\}$, $x \in \text{supp}(\eta_t)$.

Modification of the random sample P_t appears only in some stochastic strategies of global optimization e.g. Clustered Genetic Search (see Chapter 6) and Lamarcean evolution (see Section 5.3.4).

The modification may consist of *sample reduction* that removes low evaluated members (i.e. members $x \in P_t$ with the low objective value $\Phi(x)$) or removes members that lie too far from other members of the same sample P_t (with respect to the distance function in V). In extreme cases of sample reduction only the single, best evaluated member may remain. The following mapping will be useful for the best element selection.

$$b(P_t) = x;\ \Phi(x) = \max\{\Phi(\xi),\ \xi \in \text{supp}(\eta_t)\},\ P_t = (\mathcal{D}, \eta_t) \tag{2.19}$$

Another possibility of sample modification is replacing each element $x \in P_t$ by the result of some simple (low complex) local method $\text{loc}(x)$. The transformed sample may be used for the rough location of the basins of attractions in the first phase in a two-phase strategy (see Chapter 6).

The modification of the sample P_t may consist also of retaining some points from previous samples $\{P_i\}$, $i = 0, 1, \ldots, k-1$ that are located in the central parts of the basins of attractions (in sense of the problem $\mathbf{\Pi_4}$ definition).

The modified sample may replace or complement the current one in the next computation step of the stochastic strategy (see e.g. Section 5.3.4) or may be utilized only by the stopping rule and post processing of final results.

The stopping rule is one of the most difficult parts of a stochastic global optimization strategy to define and formally verify. Roughly speaking, we have to decide whether the main goal of the strategy is reached at a satisfactory degree. The simplest possibility consists of tracing the mean objective value

$$\bar{\Phi}_t = \frac{1}{\#P_t} \sum_{x \in P_t} \Phi(x) = \frac{1}{\#P_t} \sum_{y \in \text{supp}(\eta_t)} \Phi(y)\eta_t(y) \tag{2.20}$$

for consecutive samples $P_0, P_1, P_2, \ldots$. The strategy is stopped if $\bar{\Phi}_t$ exceeds the assumed threshold.

Another possibility is to use the Bayesian stopping rules that estimate the expected number of local extremes which remain to be detected (see e.g. Zieliński [208], Zieniński, Neuman [209], Betró [22], Hulin [87]).

Making the next epoch sample P_{t+1} may be obtained by some random operations performed on the previous sample P_t members (e.g. genetic operations described in Sections 3.5, 3.7) or sampled from P_t according to the probability distribution known explicitly. Some information coming from earlier epochs

$P_{t-1}, P_{t-2}, \ldots$ may also be involved, but the range of such retrospective is strictly constrained in the case of particular strategies.

No matter how the new sample is obtained, we will denote by $h_{t+1} \in \mathcal{M}(\mathcal{D})$ the probability distribution that characterized the appearance of clones of points from $\mathcal{D}$ in the multiset P_{t+1}. As we mentioned just before, h_{t+1} is computed explicitly for some algorithms (e.g. simple Monte Carlo strategies) while in other cases the sampling rule can only be established (e.g. genetic algorithms). Sometimes we are also able to compute h_{t+1} explicitly for genetic algorithms (e.g. Simple Genetic Algorithm), however, its complexity greatly exceeds the complexity of heuristic, stochastically equivalent sampling rules. The features of probability distributions $\{h_t\}$, $t = 0, 1, 2, \ldots$ determine the behavior of the whole strategy and, in particular, its efficiency.

In the case of adaptive stochastic strategies the next epoch probability distribution h_{t+1} may depend on the control parameter vector $u(t)$ that comprehends some feedback information (e.g. standard deviations of previous samples) or external controls set by user.

We will finish this section by the presentation of the Pure Random Search (PRS) strategy which is perhaps one of simplest stochastic global searches that may be used for solving problems $\mathbf{\Pi_1}$ or $\tilde{\mathbf{\Pi}}_\mathbf{1}$. It will be used further as the reference strategy due to its simplicity and good asymptotic features.

This strategy works according to the scheme shown in Figure 2.2, while:

- Initial sample P_0 is obtained by the μ-times ($\mu \in \mathbb{N}$) multiple sampling according to the uniform probability distribution over $\mathcal{D}$. Moreover, the variable $\hat{x}$ is initialized by an arbitrary point from $\mathcal{D}$.
- Sample evaluation consists of computing $\{\Phi(x)\}$, $x \in \mathrm{supp}(\eta_t)$ where η_t is the occurrence function of the sample P_t.
- Modification of the random sample consists of finding $\hat{x}_t = b(P_t)$, which is the best evaluated sample member at the epoch t. The variable $\hat{x}$ is updated $\hat{x} = \hat{x}_t$ if $\Phi(\hat{x}) \leq \Phi(\hat{x}_t)$ in the case of $\mathbf{\Pi_1}$ solution or if $\tilde{\Phi}(\hat{x}) \geq \tilde{\Phi}(\hat{x}_t)$ by solving $\tilde{\mathbf{\Pi}}_\mathbf{1}$.
- The stopping rule takes into account only the current values of $\hat{x}$ and $\Phi(\hat{x})$.
- Making the next epoch sample P_{t+1} is independent of the previous samples $P_{t-1}, P_{t-2}, \ldots$. It consists of μ-times multiple sampling according to the uniform probability distribution over $\mathcal{D}$, like at the initial step.

2.3 Basic features of stochastic algorithms of global optimization

We start with two definitions that try to formalize the ideas of asymptotic correctness and asymptotic guarantee of success mentioned in the literature (see e.g. [83], [137]). Both features are related to solving global optimization problems $\{\mathbf{\Pi}_i, \tilde{\mathbf{\Pi}}_i, \mathbf{\Pi}_i^{\mathbf{a}j}, \tilde{\mathbf{\Pi}}_i^{\mathbf{a}j}\}$, $i = 1, 2, 3; j = 1, 2, 3$ by using stochastic strategies.

Definition 2.16. *We can say that the global optimization stochastic strategy is asymptotically correct in the probabilistic sense if it finds the global maximizer (minimizer) with the probability 1 after the infinite number of epochs.* □

Definition 2.17. *We can say that the global optimization stochastic strategy has the asymptotic guarantee of success if it finds all local maximizers (minimizers) with the probability 1 after the infinite number of epochs.* □

Infinite working time which is mentioned in the above definitions 2.16, 2.17 may result in the arbitrarily large number of sample points that have to be created and evaluated. In fact, both definitions characterize not only the stochastic global optimization strategy but also the particular optimization problem to be solved. More correctly, the stochastic global optimization strategy satisfies the asymptotic correctness condition or has the asymptotic guarantee of success if the proper conditions are satisfied for the particular admissible set $\mathcal{D}$ and the objective function Φ.

Both conditions defined above do not deliver the useful criterion for stopping the stochastic strategy. The global/local minimizer may appear in the population P_t, but we have no information about its appearance. The asymptotic guarantee of success and asymptotic correctness are the only necessary conditions that have to be satisfied. In other words, the strategy that does not satisfy the asymptotic correctness never samples the global minimizers/maximizers (they lie in its "tabu" region with the constant zero sampling probability), so it is completely useless for solving problems $\mathbf{\Pi_1}, \tilde{\mathbf{\Pi}}_\mathbf{1}$. The same comment may be assigned to the asymptotic guarantee of success in the context of solving $\{\mathbf{\Pi}_i, \tilde{\mathbf{\Pi}}_i\}$, $i = 2, 3$ problems.

The next feature of the stochastic global optimization strategies that will be considered is the *global convergence* (see Rudolph [143], Beyer, Rudolph [27]). Let $x^* \in \mathcal{D}$ be the solution to the problem $\mathbf{\Pi_1}$ (x^* is one of the global maximizers of Φ), and $\Phi(x^*)$ is the maximum objective value on $\mathcal{D}$. Let us consider the random sequence

$$\{Y_t = \Phi(b(P_t))\},\ t = 0, 1, 2, \ldots \tag{2.21}$$

where b is the mapping defined by the formula 2.19 which selects the best evaluated member of the sample.

Definition 2.18. *We can say that the stochastic global optimization strategy is globally convergent if the random sequence $\{Y_t\}$, $t = 0, 1, 2, \ldots$ defined by the formula 2.21 converges completely to $\Phi(x^*)$, which means that*

$$\forall \varepsilon > 0 \left(\lim_{t \to +\infty} \sum_{j=0}^{t} \Pr\left\{ (\Phi(x^*) - Y_j) > \varepsilon \right\} \right) < +\infty$$

□

The condition defined above will be discussed extensively in Section 4.1.3. Features of stochastic global optimization strategies that will be helpful in analysis of problem $\mathbf{\Pi_4}$ are currently difficult to specify. Some technical quantities and notions like sampling measures and strategy heuristics have to be prepared earlier. They will be discussed in Section 4.1.2.

2.4 Genetic algorithms in action – solution of inverse problems in the mechanics of continua

This section presents three complex computational examples that show the skill of genetic strategies in solving continuous global optimization problems. Problems under consideration are related to the optimal design and defectoscopy of mechanical structures. The example selection was motivated in particular by their clear definitions and the possibility of impressive, graphical presentation of results.

All the presented examples show solutions of *inverse problems* (topological optimization, optimization of physical parameters of structures etc.) where each objective evaluation needs the solution of the *direct problem* which is usually given as the boundary-value problem for partial differential equations or its discrete, algebraic versions obtained by a particular numerical method (e.g. finite element method, boundary element method, see [210], [38] for details). Various genetic optimization techniques (classical evolutionary algorithm, Lamarcean evolution, Clustered Genetic Search, etc.) suitable for continuous problem solution will be applied.

Example 2.19. Topological optimization of a crane-arm truss.

This example is taken from the research paper of Burczyński, Beluch, Długosz, Kuś and Orantek [39]. The optimization problem under consideration may be classified as $\tilde{\mathbf{\Pi}}_\mathbf{1}$. It consists of finding the truss geometry e.g. the number of truss members, member cross sections and the connection topology, so that the resulting structure will be the lightest one. We assume the structure to be linearly elastic, so it has to satisfy the linear state equations

$$Ku = F \tag{2.22}$$

where K is the stiffness matrix of the truss, u and F stand for the displacement vector and the external force vector at the truss joints respectively (we refer to Zienkiewicz, Taylor [210] for details). Moreover, the following constraints have to be satisfied:

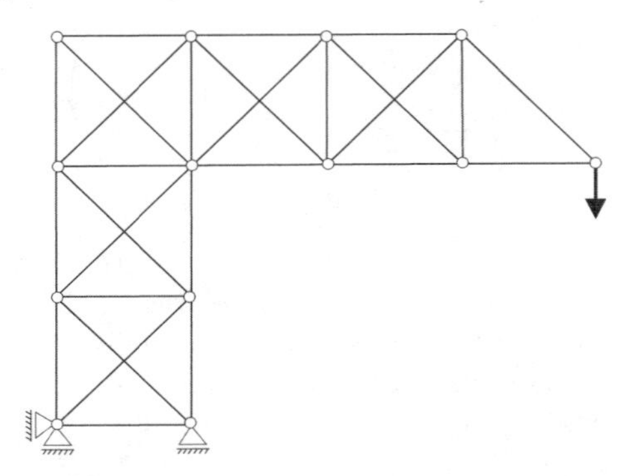

Fig. 2.3. Initial shape of the truss.

- stresses in truss member and displacement at each truss joint have to be less than admissible ones,
- normal forces have to be less than bulking ones.

Decision variables are composed of three groups that characterize:

1. information about existing truss members,
2. information about areas and cross sections of truss members,
3. coordinate of free truss joints.

All three groups of decision variables encoded as a real-valued vector constitute the genotype of the individual (see Section 3.1). A constant size population (random sample) of 5000 individuals was processed. Rank selection (see Section 3.4.4), classical phenotypic mutation (see Section 3.7.1) and arithmetic crossover (see Section 3.7.2) were utilized in order to make new populations in each epoch.

After about 1500 epochs the population was stabilized (no significant improvement of the best fitted individual was observed), so the further processing of up to 2000 epochs did not bring the essential improvement. The total weight of the crane-arm truss decreased from the initial 5497,44 kg to the final one 3406,07 kg. The transformation of the truss topology is shown in Figures 2.3, 2.4.

Example 2.20. Identification of internal defects in lumped elastic structures.

The genetic computations presented below are applied to the detection of undesirable voids (cracks and lumped voids) inside the massive (lumped), elastic structures. The problem is formulated as a global optimization one, where the control variables determine the number, shapes and locations of voids. Decision variables may be transformed unambiguously to the physical parameter distribution inside the structure body. The structure was subjected to cyclic loading, then external strains and eigenfrequencies were measured.

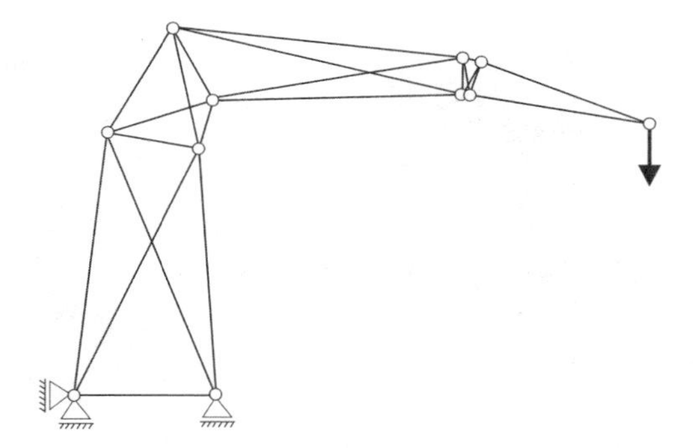

Fig. 2.4. Final shape of the truss after topological optimization.

Genetic optimization allows us to find such decision variables that get the closest simulated response of the structure to the real, measured one.

The example description follows the research paper written by Burczyński, Kuś, Nowakowski and Orantek [40].

Let us assume that, for all the structures analyzed further, the elastodynamic partial differential equations were satisfied:

$$\begin{aligned} &\mathrm{div}\sigma - \rho\ddot{u} = 0, \text{ in } S \times [0,T] \\ &\sigma = C\varepsilon, \; \varepsilon = \tfrac{1}{2}(\nabla u + \nabla u^T) \end{aligned} \tag{2.23}$$

where u is the displacement field and $\ddot{u}$ its second partial derivative with respect to the time variable, σ and ε are second order tensor fields of stress and strain respectively, C stands for the 4th order tensor field of the elastic constitutive law, finally ρ is the material density field (see e.g. Derski [58]). In the above formulas S denotes the structure domain and $[0,T]$ the time interval in which we are looking for the solution. The symbols div and ∇ denote the divergence computed from the tensor field and the gradient operator computed separately for each coordinate of the vector-valued function respectively.

The proper boundary conditions were satisfied on the external boundary of the structure $\varGamma$ and on the internal boundary of defects. The adequate initial condition is also satisfied inside the structure S at $t = 0$.

If the body S undergoes free vibration the governing equation 2.23 takes the form:

$$\mathrm{div}\sigma - \omega^2 \rho u = 0, \text{ in } S \times [0,T] \tag{2.24}$$

where ω denotes the circular eigenfrequency of the structure.

The objective function $\tilde{\varPhi}$ will penalize discrepancies between the observed displacement $\hat{u}$ and consecutive circular eigenfrequencies $\hat{\omega}_i$ and simulated values of such quantities u, ω_i. The objective is given as the linear combination:

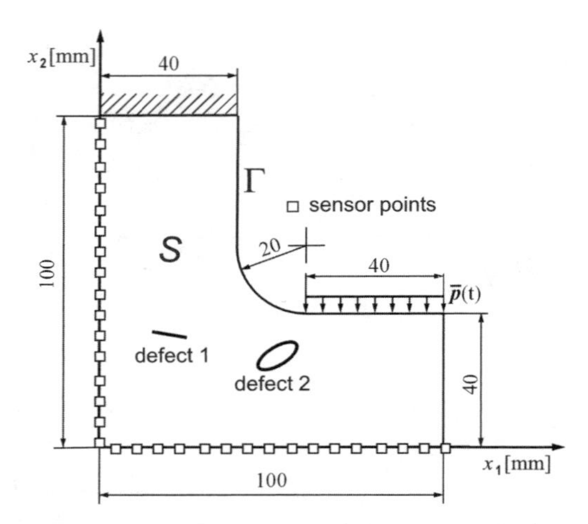

Fig. 2.5. Loading of the plate at the defect identification problem.

$$
\begin{aligned}
\tilde{\Phi} &= w_1\tilde{\Phi}_1 + w_2\tilde{\Phi}_2 \\
\tilde{\Phi}_1 &= 0.5\sum(\hat{\omega}_i - \omega_i)^2 \\
\tilde{\Phi}_2 &= 0.5\int_0^T \int_\Gamma (\hat{u} - u)^2 d\sigma dt
\end{aligned}
\tag{2.25}
$$

where w_1, w_2 are the proper numeric weights.

The global optimization problem of type $\tilde{\mathbf{\Pi}}_1$ was solved. It consists of finding the global minimizers of $\tilde{\Phi}$ with respect to u, ω_i. However u, ω_i are not the target decision variables of the identification problem. The target ones are the number of internal defects and their shape parameters. The dependency between the target decision variables and variables u, ω_i as well as the objective function $\tilde{\Phi}$ and its first and second variables $D\tilde{\Phi}$, $D^2\tilde{\Phi}$ is established by solving the system 2.23, 2.24 by the boundary element method (see Burczyński [38]). As usual in the case of identification problems the objective value is zero in each global minimizer.

Genotypes are real-valued vectors that encode defect shape parameters. Parameter vectors take their values from some brick $\mathcal{D} \subset \mathbb{R}^N$ which is the admissible set of solutions. The evolutionary algorithm utilized for global optimization problems applies the tournament and rank selections (see Sections 3.4.3 and 3.4.4), phenotypic mutation and arithmetic crossover (see Sections 3.7.1 and 3.7.2) and the special genetic operation called "gradient mutation". Detailed description of the last operation may be found in Section 5.3.4.

The first computations were performed for the plate (see Figure 2.5). The defectoscopy problem consists of identification of 3 defects (0, 1, 2 or 3 defects) of the elliptical shape. Each defect is parametrized by five real numbers: two center coordinates, two half-axis lengths and a measure of the angle between the axis $0x_1$ and the large half-axis. Only the first three eigenfrequencies

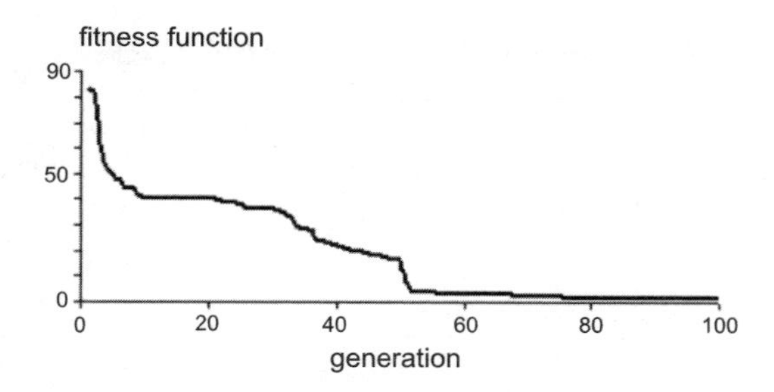

Fig. 2.6. Fitness of the best individual in consecutive epochs by defect identification in plate.

$\omega_1, \omega_2, \omega_3$ were taken into account. The plate was loaded by vibratory, periodical boundary loading $p(t) = p_0 \sin(\omega t)$ with the frequency $\omega = 15708\,\mathrm{rad/sec}$ and the amplitude $p_0 = 40\,\mathrm{kN/m}$. The vibration impulse was activated in the time period $t \in [0, 600\,\mu s]$. The plate material was homogeneous, isotropic and linearly elastic with a Young modulus 0.2×10^{12} Pa, Poisson ratio 0.3 and density $\rho = 7800\,\mathrm{kg/m^3}$. The left and lower sides of the plate were instrumented with 64 sensors that measured $\hat{u}$ and $\hat{\omega}_i$, $i = 1, 2, 3$.

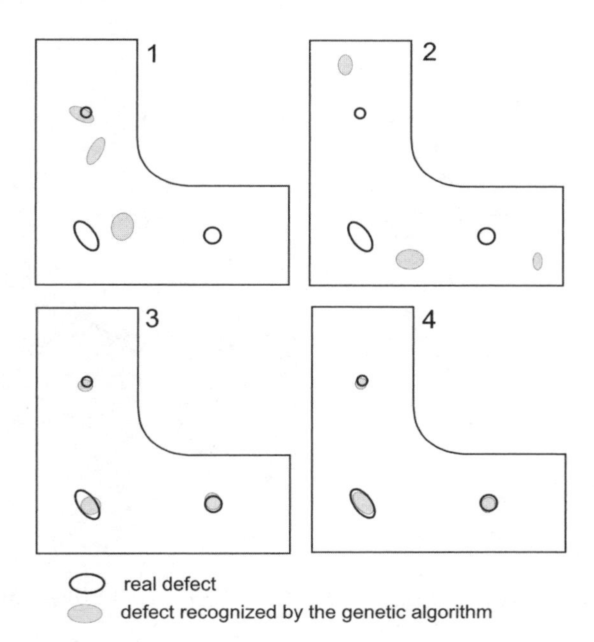

Fig. 2.7. The defects identification by the best individual phenotype in consecutive epochs: 1 - 1th, 2 - 10th, 3 - 50th, 4 - 100th.

The genotype of each individual was composed of 15 coordinates (5 coordinates for each of three defects). The population contains 3000 individuals at each epoch. Figure 2.6 shows the objective value of the best fitted individual in the consecutive genetic epochs. The next Figure 2.7 delivers the snapshots from the identification process at its four selected phases.

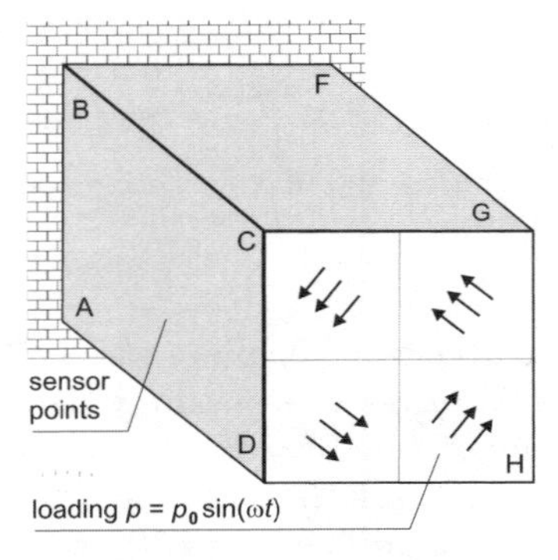

Fig. 2.8. Loading and the sensor location on the defected rod.

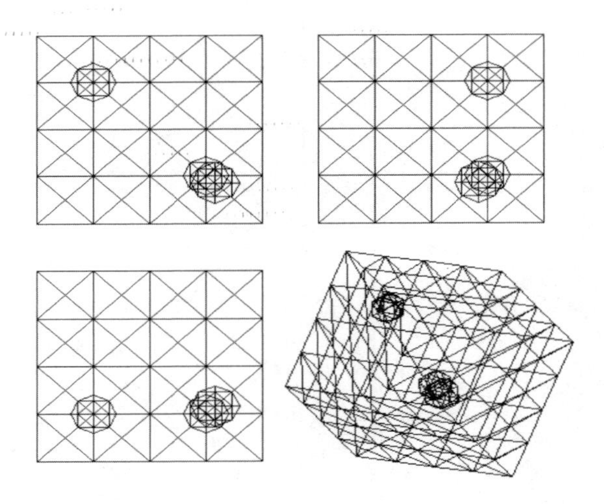

Fig. 2.9. Location of defects identified in road versus real life ones after 200 epochs.

The next computations were performed for a cubic rod with an edge length 0.2 m (see Figure 2.8). The rod is stiffly fastened at its back wall, and periodically loaded at its front wall. The loading has the norm uniformly distributed in the whole wall area and the direction different in each quarter. The norm of the loading depends on time according to the formula $p(t) = p_0 \sin(\omega t)$ with the frequency $\omega = 31\,\text{rad/sec}$ and the amplitude $p_0 = 15000\,\text{kN/m}$. The loading causes the periodical twisting deformation of the rod. The mass density of the rod is $100\,\text{kg/m}^3$, the shear modulus 10^6 Pa and the Poisson ratio 0.25.

The rod is instrumented by 64 sensors uniformly located in the four back walls. The structure contains two spherical defects, each described by four real valued parameters (three center coordinates and radius length). As a consequence, the genotype is composed of 8 coordinates (four coordinates for a single defect).

The shape and the location of the identified voids after 2000 epochs were shown in Figure 2.9. The results mentioned were obtained for exact values of $\hat{u}$ and $\hat{\omega}_i$, $i = 1, 2, 3$. Additional tests that assume randomly perturbed values of measured displacements and eigenfrequencies did not bring a significant decrement in the void identification accuracy. This strategy using "gradient mutation" is considerably less computationally complex than the simple evolutionary one and makes the results more accurate.

Example 2.21. Optimal pretraction design in the simple cable structure.

The global optimization problem of type $\tilde{\mathbf{\Pi}}_4$ is considered. It consists of finding the central parts of the basins of attraction of the objective $\tilde{\Phi}$ which mainly expresses the energy of internal strains of the simple cable structure (hanging roof). The recognition of the basins of attraction was the first phase of the two-phase stochastic global optimization strategy (see Chapter 6). In the second phase fast local optimization methods were started separately in each basin finding accurate approximations of local minimizers. Additionally, parameters such as the basin diameter and "depth" (the objective variation inside the basin) may be helpful in the sensitivity analysis of the obtained minimizers.

The presented results were taken from research papers written by Telega [184] and Telega, Schaefer [186].

The structure under consideration is composed of unconnected cables stretched in two perpendicular directions lying in a single horizontal plane. Cables are fastened at their ends to the stiff square frame of the area $S = [0,1]^2 \subset \mathbb{R}^2$ in such a way that the frame sides are parallel or perpendicular to each cable. The resulting network structure is loaded by forces perpendicular to the frame plane. The loading is characterized by its surface intensity q.

The linearized and homogenized state equation for such a structure is the following (see Cabib, Davini, Chong-Quing Ru [42]):

$$\begin{cases} -\text{div}(\sigma D u) = q & \text{in} \quad S \\ \qquad u = 0 & \text{on} \quad \partial S \end{cases} \tag{2.26}$$

where ∂S stands for the frame contour and u for the cable's perpendicular displacement field. The strain tensor is given by the formula:

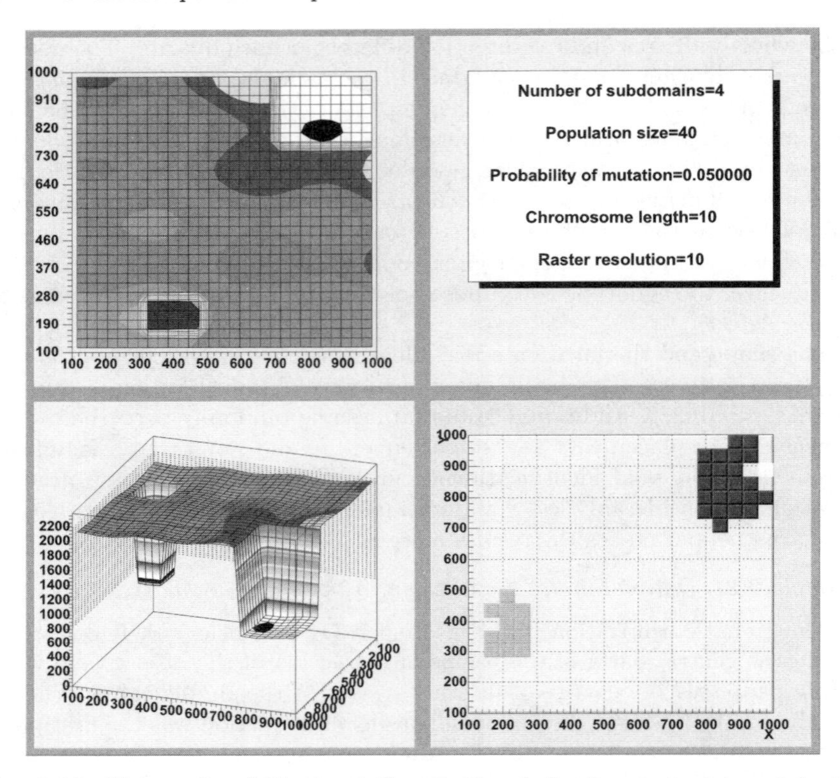

Fig. 2.10. The results of Clustered Genetic Search for the strategy tuned for two noticeable local minimizers.

$$\sigma(x) = \mathrm{diag}(\sigma_1(x_2), \sigma_2(x_1)) \tag{2.27}$$

where $\sigma_i(x_j)$ denotes the pretraction displacement of cables parallel to the axis $0x_i$ which depends only on the variable x_j, for $i, j = 1, 2$.

The objective function is given by:

$$\tilde{\Phi} = \int_S \sigma \left| u_\sigma \right|^2 dx + P(\sigma) \tag{2.28}$$

where u_σ denotes the cable displacement obtained by pretractions σ_1, σ_2. The function P denotes the middle penalty function that penalizes the deviations from the typical cable parameters and turns back the zero value for parameters offered by manufacturers. It was one of the reasons that the objective function is multimodal.

Pretractions σ_1, σ_2 and loading intensity q satisfy the constraints:

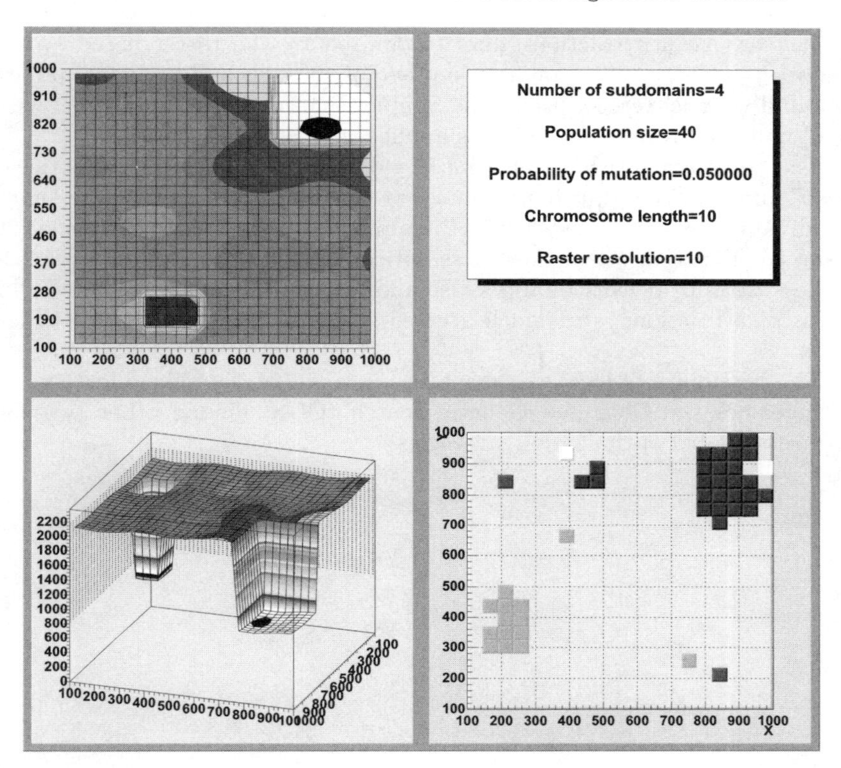

Fig. 2.11. The results of Clustered Genetic Search for the strategy tuned for less noticeable local minimizers.

$$
\begin{gathered}
0 < \lambda < \sigma_1, \sigma_2 < \Lambda, \ \lambda, \Lambda \in \mathbb{R} \\
\int_S (\sigma_1 + \sigma_2)\, dx \in [2\lambda, 2\Lambda] \\
\int_S q\, dx = 0
\end{gathered}
\tag{2.29}
$$

Using some additional assumptions (see [42]) the minimization problem for $\tilde{\Phi}$ and pretractions σ_1, σ_2 satisfying them 2.29 have more than one global minimizer.

The 10×10 cable network was assumed for computations. The loading q is displaced centrally in the 6×6 segment of cables. The pretraction's constraints were assumed as $\lambda = 10^2\,\text{kN}, \ \Lambda = 10^3\,\text{kN}$. Computational results were only presented for the two cables closest to the frame ∂S because of the double symmetry of the problem. The admissible set was $\mathcal{D} = [\lambda, \Lambda]^2$ in this case. Figure 2.10 shows the graph of the function $\tilde{\Phi}$ obtained by the multiple solving of the direct problem 2.26. The objective function has two noticeable "deep" minima with large basins of attraction and more "shallow" local minimizers (there is a small objective variation inside the basin of attraction).

The basins of attractions were recognized by the raster based strategy described in Chapter 6. A raster composed of 400 cells and a small population of 40 individuals were utilized. The Simple Genetic Algorithm (SGA) with a small mutation rate $p_m = 0.05$ was applied in the global phase.

The genetic algorithm well tuned (see definition 4.63) to the noticeable "deep" minima allowed us to recognize two basins of attraction by the analysis of individual density in raster cells. Simultaneously, the strategy filters out the basins of attraction of "shallow", less noticeable minimizers (see Figure 2.10).

Less rigorous density analysis also allowed us to recognize some "shallow" basins of attractions with much greater values of local minima (see Figure 2.11).

The possibility of local extreme filtering is one of the unique features that distinguishes the Clustered Genetic Search (CGS) among other two-phase stochastic global optimization strategies.

3

Basic models of genetic computations

In this chapter we would like to present the basic objectives and operations that are used to build the genetic optimization model. We start with the fundamental encoding operation that transforms the original global optimization problem to the domain of genetic codes, called the genetic universum. We will enrich the relation between the admissible domain and the genetic universum with the decoding operation, which will be especially helpful in the case of searching in continuous domains. Next, the basic group of operations (selections, mutations and crossovers) will be discussed as methods of random transformation of the population to the next epoch. Special emphasis is put on features of binary genetic operations (operations dealing with binary genetic code) that will be intensively used in the next chapter. The detailed model of the Simple Genetic Algorithm (SGA) is also defined. Schwefel's taxonomy of evolutionary single- and multi-population (multi-deme) computation models is presented at the end of this chapter.

3.1 Encoding and inverse encoding

One of the basic mechanisms used in genetic computations is the special method of representation of random sample members, that are called here *individuals*. This representation is obtained by the proper encoding of some points from the admissible domain $\mathcal{D}$. There are at least two goals of individual encoding in genetic computations:

1. Selection of the arbitrary subset of admissible points from $\mathcal{D}$ which can be effectively checked during the optimization process.
2. The transformation of design variables that enables the special stochastic operations (genetic operations), that can replace sampling procedure, to be performed.

Let us introduce the necessary notation:

R. Schaefer: *Foundation of Global Genetic Optimization*, Studies in Computational Intelligence (SCI) **74**, 31–53 (2007)
www.springerlink.com

$\mathcal{D}_r \subseteq \mathcal{D}$ Set of points in the admissible domain which constitute the grid that will be effectively checked. Elements of $\mathcal{D}_r$ will be called *phenotypes.*

$r = \#\mathcal{D}_r$ Finite or infinite cardinality of the set of phenotypes.

U Genetic universum as the whole set of genetic codes used for marking phenotypes. Members of the genetic universum will be called *genotypes.* We sometimes assume that the genetic universum is closed with respect to the set of genetic operations.

The necessary coherency condition among cardinalities of the above sets has to be satisfied:

$$\#U = r = \#\mathcal{D}_r \tag{3.1}$$

Please note that the above condition holds for either infinite sets of phenotypes or genetic universum. The variable r will be a proper cardinal number in this case.

Definition 3.1. *Encoding will be one-to-one mapping (bijection)*

$$\text{code} : U \rightarrow \mathcal{D}_r$$

□

We can try to define the new mapping, which assigns to the genotype $x \in U$ admissible points from $\mathcal{D} \setminus \mathcal{D}_r$ lying close to the point $\text{code}(x) \in \mathcal{D}$.

Definition 3.2. *The inverse encoding (decoding) associated with the encoding function* $\text{code} : U \rightarrow \mathcal{D}_r$ *is the partial mapping, which is not necessarily defined for all points in* $\mathcal{D}$ *(* $\text{Dom}(\text{dcode}) \subset \mathcal{D}$ *where the operator* Dom *turns back the function domain):*

$$\text{dcode} : \mathcal{D} \longmapsto U$$

that satisfies the following conditions:

1. $\text{dcode}(\text{code}(x)) = x, \ \forall x \in U$
2. $\forall x \in U, \ \forall y \in \mathcal{D} \setminus \mathcal{D}_r, \ \forall z \in U \setminus \{x\}, \ d(y, \text{code}(x)) < d(y, \text{code}(z)) \Rightarrow \text{dcode}(y) = x$*, where* d *is the metric in the space* V*.* □

Remark 3.3. The partial function dcode satisfies the following conditions:

1. It is not defined for the admissible points that lie in the middle of two or more neighboring phenotypes, in particular for $x \in \mathcal{D} \setminus \mathcal{D}_r$ so that $\exists \chi, \ \lambda \in \mathcal{D}_r; \ d(x,\chi) = d(x,\lambda)$ and there is no other phenotype $\gamma \in \mathcal{D}_r$ in such a way that $\gamma \neq \chi, \ \gamma \neq \lambda$ and $d(x,\gamma) < d(x,\chi), \ d(x,\gamma) < d(x,\lambda)$.

2. It is always onto (is surjective).
3. The restriction dcode $|\mathcal{D}_r : \mathcal{D}_r \to U$ is a one-to-one function as the inverse of encoding function code.
4. It can be arbitrarily extended to the function defined on the whole admissible domain $\mathcal{D} \to U$. One possible solution is to follow the lexicographic total order "$\succ$" which may be introduced in $\mathcal{D}_r$ ($\forall \chi,\ \lambda \in \mathcal{D}_r\ \chi \succ \lambda \Leftrightarrow \chi_i \geq \lambda_i,\ i = 1, \ldots, N, \exists k \in \{1, \ldots, N\};\ \chi_k > \lambda_k$). If $x \in \mathcal{D} \setminus \mathcal{D}_r$ and $\chi,\ \lambda \in \mathcal{D}_r$ are such as in the first item, then dcode(x) may be set dcode(λ) if $\chi \succ \lambda$ or dcode(χ) otherwise. Such an extension is also onto. □

Remark 3.4. The most interesting for further consideration will be the case for which $\#\mathcal{D}_r < +\infty$, and

$$\begin{aligned} &\text{meas}(\text{Dom}(\text{dcode})) = \text{meas}(\mathcal{D}) \\ &\forall x \in U\ \text{meas}(\text{dcode}^{-1}(x)) > 0 \end{aligned} \tag{3.2}$$

which also implies $\text{meas}(\mathcal{D}) = \sum_{x \in U} \text{meas}(\text{dcode}^{-1}(x))$. The above conditions may be satisfied if $\mathcal{D}$ is convex and sufficiently regular, so that the Voronoi Tasselation can be successfully performed for the finite set of phenotypes $\mathcal{D}_r$ in $\mathcal{D}$ (see e.g. [131]). □

Remark 3.5. In some special cases U will be equipped with the metric function $d_U : U \times U \to \mathbb{R}_+$. Arabas (see [5], Section 4.2) calls for another condition for well-conditioned encoding functions:

$$\begin{aligned} &\forall x_1, x_2, x_3 \in \mathcal{D}_r\ \ d(x_1, x_2) \geq d(x_1, x_3) \Rightarrow \\ &d_U(\text{dcode}(x_1), \text{dcode}(x_2)) \geq d_U(\text{dcode}(x_1), \text{dcode}(x_3)) \end{aligned} \tag{3.3}$$

This condition would prevent the creation of the artificial local extrema in the genotype domain due to the poor topological condition of encoding mapping. □

The probabilistic measures $\theta \in \mathcal{M}(\mathcal{D}_r)$ play a crucial rule in the process of genetic sampling intensively discussed in the following sections. Each measure of such a type induces another measure $\theta' \in \mathcal{M}(\mathcal{D})$ which is concentrated on the set $\mathcal{D}_r \subset \mathcal{D}$.

Such simple extensions have some serious disadvantages. In particular, if the phenotype set is discrete $\mathcal{D}_r < +\infty$, then it is possible to get the measurable set $A \subset \mathcal{D}$ so that $A \cap \mathcal{D}_r = \emptyset$ and then $\theta'(A) = 0$, while its Lesbegue measure will be strictly positive ($\text{meas}(A) > 0$) and arbitrarily close to $\text{meas}(\mathcal{D})$.

In order to avoid this problem we will introduce a new measure unambiguously induced by the discrete measure θ. If the condition 3.2 holds, then perhaps the simplest option is to take the measure with the density defined by using dcode mapping:

$$\rho_\theta(y) = \frac{\theta(\text{code}(z))}{\text{meas}(\text{dcode}^{-1}(z))} \quad \text{if} \;\; y \in \text{dcode}^{-1}(z) \tag{3.4}$$

It is easy to see that ρ_θ is a piecewise constant function defined almost everywhere in $\mathcal{D}$. "Almost everywhere in $\mathcal{D}$" means for all points in $\mathcal{D}$ possibly excluding points that belong to some subset of a zero Lesbegue measure.

Of course, assuming any strictly positive discrete measure on phenotypes $\theta \in \mathcal{M}(\mathcal{D}_r)$; $\theta(y) > 0 \; \forall y \in \mathcal{D}_r$ for all sets $A \subset \mathcal{D}$ the condition $\text{meas}(A) > 0$ always implies $\int_A \rho_\theta dx > 0$. Moreover, $\text{meas}(A) = 0$ holds only if $\int_A \rho_\theta dx = 0$.

3.1.1 Binary affine encoding

This is perhaps the oldest and most traditional encoding technique, introduced by Holland in his first well known work [85]. Its application to the global search in continuous multidimensional domains $\mathcal{D}$ needs a much more formal description than currently found in the literature.

The genetic universum U is traditionally denoted by Ω in this case and it is composed of all the binary strings of the finite, prescribed constant length $l \in \mathbb{N}$.

$$\Omega = \{(a_0, a_1, \ldots, a_{l-1}); a_i \in \{0,1\}, i = 0, 1, \ldots, l-1\} \tag{3.5}$$

Although binary genotypes from Ω are usually associated with chromosomes and are often used to label various discrete structures, it is easy to identify them with integers from the range $[1, 2, \ldots, r-1]$. Each binary string $i \in \Omega$ may be treated as the integer that equals its numerical value and belongs to the above range. The value $r - 1 = 2^l - 1$ constitutes the upper value of this range, then we have:

$$\#\Omega = r = 2^l < +\infty \tag{3.6}$$

The binary genetic universum Ω may also be identified with the l – times Cartesian product

$$\underbrace{Z_2 \times Z_2 \times, \ldots, \times Z_2}_{l-\text{times}} \tag{3.7}$$

where Z_2 is the group with modulo 2 addition defined on the set $\{0,1\}$. The object given by 3.7 also makes a group with the component-wise modulo 2 addition, denoted by $\oplus$, defined on the set $\{0,1\}^l$.

Let us restrict ourselves to the case in which $V = \hat{V} = \mathbb{R}^N$. We assume moreover, that the admissible set $\mathcal{D}$ is convex and contains a brick

$$\prod_{i=1}^{N} [u_i, v_i] \subset \mathcal{D} \tag{3.8}$$

being the multiple Cartesian product of intervals $[u_1, v_1], [u_2, v_2], \ldots, [u_N, v_N]$.

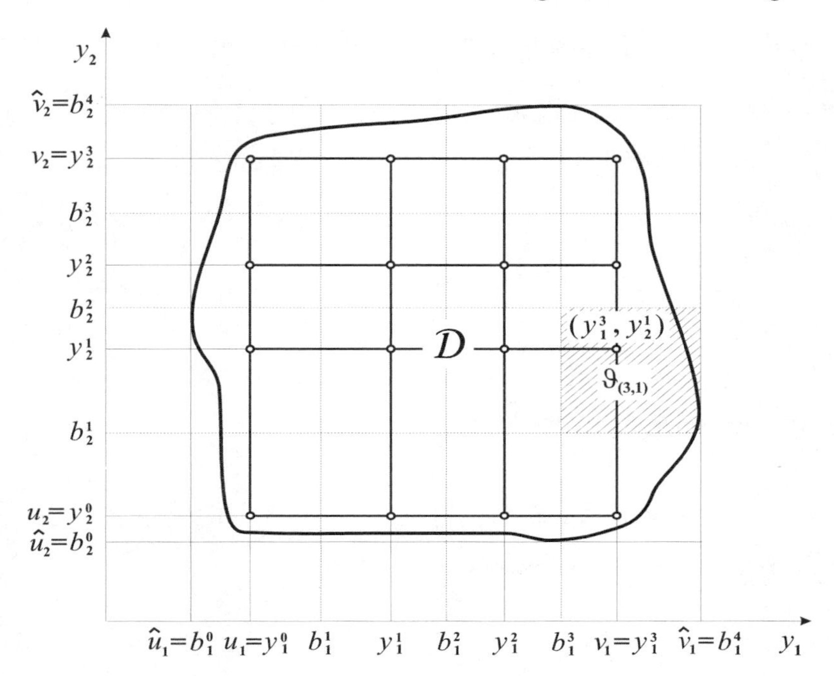

Fig. 3.1. Binary affine encoding in two dimensions

Let $\hat{u}_i, \hat{v}_i \in \mathbb{R};\ \hat{u}_i < \hat{v}_i < +\infty,\ i = 1, \ldots, N$ be the lower and upper bounds for the coordinates of admissible points i.e. $\forall y = (y_1, \ldots, y_N) \in \mathcal{D}\ \hat{u}_i \leq y_i \leq \hat{v}_i,\ i = 1, \ldots, N$.

The affine encoding mapping code_a will be defined in the following steps:

1. We select numbers $l_1, l_2, \ldots, l_N \in \mathbb{N}$ so that $\sum_{i=1}^{N} l_i = l$.
2. We define the phenotype set in the form

$$\mathcal{D}_r = \Big\{ \Big(y_1^{j_1}, \ldots, y_N^{j_N} \Big);\ y_i^0 = u_i, y_i^{2^{l_i}-1} = v_i, y_i^{j_i} < y_i^{j_i+1}, \\ j_i \in \{0, \ldots, 2^{l_i} - 2\},\ i = 1, \ldots, N \Big\}. \tag{3.9}$$

3. Each genotype $x \in \Omega$ may be represented in two equivalent forms

$$x = (a_{1,1}, \ldots, a_{1,l_1}, a_{2,1}, \ldots, a_{2,l_2}, \ldots, a_{N,1}, \ldots, a_{N,l_N}),\ a_{i,j} \in \{0, 1\}, \\ x = (j_1, \ldots, j_N),\ j_i \in [0, \ldots, 2^{l_i} - 1] \tag{3.10}$$

where $(a_{i,1}, \ldots, a_{i,l_i})$ stands for the binary code of the integer j_i.

4. Using both of the above representations of the genotype $x \in \Omega$ we may write

$$\begin{aligned} \text{code}_a(x) &= \text{code}_a(a_{1,1}, \ldots, a_{1,l_1}, a_{2,1}, \ldots, a_{2,l_2}, \ldots, a_{N,1}, \ldots, a_{N,l_N}) \\ &= \left(y_1^{j_1}, \ldots, y_N^{j_N}\right) \end{aligned} \tag{3.11}$$

where

$$j_i = \sum_{k=0}^{l_i-1} a_{i,(l_i-k)} 2^k. \tag{3.12}$$

The inverse binary affine encoding dcode_a may be obtained as follows:

1. We create the set of numbers

$$\begin{aligned} &\{b_i^s\},\ s = 0, \ldots, 2^{l_i},\ i = 1, \ldots, N, \\ b_i^0 = \hat{u}_i,\ b_i^j = \tfrac{1}{2}&\left(y_i^j + y_i^{j-1}\right),\ j = 1, \ldots, 2^{l_i} - 1,\ b_i^{2^{l_i}} = \hat{v}_i. \end{aligned} \tag{3.13}$$

2. Next we define the family of open bricks

$$\left\{\vartheta_{(j_1,\ldots,j_N)}\right\}, (j_1, \ldots, j_N) \in \Omega;\ \vartheta_{(j_1,\ldots,j_N)} = \prod_{i=1}^{N} \left(b_i^{j_i-1}, b_i^{j_i}\right). \tag{3.14}$$

3. $\text{dcode}_a(y) = (j_1, \ldots, j_N) \in \Omega \Leftrightarrow y \in \mathcal{D} \cap \vartheta_{(j_1,\ldots,j_N)}$.

It is easy to see that the above mapping is not defined on hyperplanes separating neighboring bricks 3.14. The above definitions of code_a and dcode_a operations follow the paper Schaefer, Jabłoński [154].

Remark 3.6. The construction of both operations code_a and dcode_a presented above satisfies the definitions 3.1, 3.2. In particular:

1. $\mathcal{D}_r \subset \mathcal{D}$ and code_a is one-to-one which is the effect of the phenotype set construction (see 3.9) and unambiguity of genotype representations 3.10.
2. $\forall x \in \Omega\ \text{dcode}_a(\text{code}_a(x)) = x$, because $x \in \vartheta_x$.
3. The set $\vartheta_x \cap \mathcal{D}$ constitutes the Voronoi neighborhood for $\text{code}_a(x)$ in $\mathcal{D}$ (see Preprata, Shamos [131]).
4. $\text{meas}\left(\bigcup_{x\in\Omega}(\vartheta_x \cap \mathcal{D})\right) = \text{meas}(\mathcal{D})$ and $\text{meas}(\text{dcode}_a^{-1}(x)) = \text{meas}(\vartheta_x \cap \mathcal{D}) > 0\ \forall x \in \Omega$ which prove conditions 3.2 postulated in Remark 3.4. The above conditions also enable us to effectively define the density function ρ_θ for the arbitrary discrete measure $\theta \in \mathcal{M}(\mathcal{D}_r)$, according to the formula 3.4. This construction will be discussed and applied in Section 4.1.2. □

The construction of the affine binary encoding may be partially extended by using the curvilinear coordinates to parametrize the admissible domain. The second condition postulated in definition 3.2 may be dropped in the case of non convex domains.

3.1.2 Gray encoding

Gray encoding may be understood as another case of the affine encoding described in Section 3.1.1.

Let us define the mapping $\upsilon : \Omega \to \Omega$ so that:

$$\begin{aligned} \upsilon(a) = b, \; a = (a_0, a_1, \ldots, a_{l-1}), \; b = (b_0, b_1, \ldots, b_{l-1}) \\ \Leftrightarrow \; b_i = \oplus_{j=0}^{i-1} a_j, \; i = \{0, 1, \ldots, l-1\} \end{aligned} \tag{3.15}$$

It is easy to check that υ is one-to-one (bijection) mapping, and its inverse υ^{-1} is given by:

$$\begin{aligned} \upsilon^{-1}(b) = a, \; a = (a_0, a_1, \ldots, a_{l-1}), \; b = (b_0, b_1, \ldots, b_{l-1}) \\ \Leftrightarrow \; a_i = \begin{cases} b_i & i = 0 \\ b_{i-1} \oplus b_i & i > 0 \end{cases} \end{aligned} \tag{3.16}$$

where $\oplus$ denotes the modulo 2 addition (see e.g. [10]).

Let us now introduce the Hamming metrics in the space of binary genotypes

Definition 3.7. *Hamming metrics on Ω will be given by the mapping $d_H : \Omega \times \Omega \to \{0, 1, \ldots, l\}$ such that for the arbitrary genotypes $a = (a_0, a_1, \ldots, a_{l-1})$, $b = (b_0, b_1, \ldots, b_{l-1})$ we have:*

$$d_H(a, b) = \sum_{i=0}^{l-1} a_i \oplus b_i$$

□

It will also be easy to check that:

Remark 3.8. Let $j_1, j_2 \in \{0, 1, , \ldots, 2^l - 1\}$ be two integers and $a^1, a^2 \in \Omega$ their standard binary representation, then the following implication holds:

$$|j_1 - j_2| = 1 \Rightarrow d_H(\upsilon(a^1), \upsilon(a^2)) = 1.$$

□

Gray encoding mapping will be given by the composition:

$$\text{code}_G : \Omega \to \mathcal{D}; \; \text{code}_G = \text{code}_a \circ \upsilon. \tag{3.17}$$

The inverse Gray encoding is a partial function which will also be given by the composition:

$$\text{dcode}_G : \mathcal{D} \mapsto \Omega; \; \text{dcode}_G = \upsilon^{-1} \circ \text{dcode}_a. \tag{3.18}$$

Both mappings code_G and dcode_G satisfy all the conditions imposed for binary affine encoding and inverse binary affine encoding in Remark 3.6.

The Gray encoding may be helpful in "dulling" discrepancies between the Hamming distance of two genotypes vs distance of respecting phenotypes (frequently called "Hamming cliffs") that appear by the affine binary encoding. This feature makes Gray encoding more closely satisfy the condition 3.3, which is helpful when solving continuous problems.

3.1.3 Phenotypic encoding

This is the trivial case of individual representation applied in many instances of evolutionary algorithms (EA). In order to define this type of coding we distinguish the affine isometry between $\mathbb{R}^N$ and the search space $I : \mathbb{R}^N \to V$. If $V = \mathbb{R}^N$ then I will be the identity mapping. The genetic universum will be given by:

$$U = I^{-1}(\mathcal{D}) \tag{3.19}$$

The role of the encoding operator will be played by $I|U$ and inverse coding by $I^{-1}|\mathcal{D}$. Please note that I is bijection (one-to-one mapping between $\mathbb{R}^N$ and V) and $I|U$ is the bijection among U and $\mathcal{D}$.

The above mappings satisfy all the conditions demanded by the definitions 3.1, 3.2. In most instances of evolutionary algorithms studied later we will identify U with $\mathcal{D}$. The important difference with respect to the previous, binary case is that the genetic universum (set of genetic codes) is infinite and uncountable (its cardinality is continuous) (see e.g. Schwartz [162]).

3.2 Objective and fitness

During the genetic optimization process crucial operations are performed on the individual codes, the elements of the genetic universum U (or Ω in case of binary encoding). As a natural consequence, the objective function that evaluates potential solutions has to be transported in some way to the new domain U. The new function $f : U \to [0, M]$ traditionally called *fitness function*, that evaluates genotypes has to be strongly correlated with the objective $\Phi : \mathcal{D} \to [0, M]$. Perhaps the simplest possibility is to define:

$$f(x) = \Phi(\text{code}(x)) \ \forall x \in U \tag{3.20}$$

Sometimes, we would like to change selection pressure with respect to the worst fitted individuals. Such a feature may be obtained by the linear or non-linear scaling of fitness with respect to the original objective value by using strongly monotone "scaling" function (see e.g. Goldberg [74], Bäck, Fogel, Michalewicz [15], [10]):

$$Scale : [0, M] \rightarrow [0, M'] \tag{3.21}$$

where $M' < +\infty$ is the new upper bound for the individual evaluation. Using the scaling function we obtain:

$$f(x) = Scale(\Phi(\text{code}(x))) \; \forall x \in U \tag{3.22}$$

In the next parts of this book we do not distinguish between constants M and M' treating them as the single, generic constant M.

Other, more sophisticated ways to define the fitness mapping will be described in Sections 5.3.4, 5.3.5.

3.3 The individual and population models

Genetic algorithms generally fall into the schema of stochastic population search described in Section 2.2 and roughly presented in Figure 2.2. Because genetic algorithms always represent design variables from the admissible set $\mathcal{D}$ (exactly from its subset $\mathcal{D}_r$) as the genetic codes U by using encoding function (see Section 3.1) we will consider populations $\{P_t\}$, $t = 0, 1, \ldots$ as multisets of clones from the genetic universum U rather than from the set of phenotypes $\mathcal{D}_r$.

Population members in populations processed by genetic algorithms are called individuals. The consecutive iterations of genetic algorithm indexed by $t = 0, 1, 2, \ldots$ will be called *genetic epochs* or simpler epochs.

According to the definition 2.8 we will represent P_t as a pair (U, η_t), $t = 0, 1, \ldots$ where η_t stands for the occurrence function in the tth genetic epoch. If the population P_t is finite and distinction among clones of the same genotype is necessary we will represent it in the set-like form postulated in Remark 2.15.

Summing up, the individual x, being a member of the population P_t in the arbitrary genetic epoch t, is characterized by the triple:

$$\begin{array}{ll} x \in \text{supp}(\eta_t) \subset U & \text{the individual's genotype} \\ \text{code}(x) \in \mathcal{D}_r \subset \mathcal{D} & \text{the individual's phenotype} \\ f(\text{code}(x)) \in [0, M] \subset \mathbb{R}_+ & \text{the individual's fitness} \end{array} \tag{3.23}$$

the set $\text{supp}(\eta_t)$ stands for the *genetic material of the population* P_t.

In the case of more extended genetic mechanisms (see e.g. Beyer [26]) discussed in Chapter 5 the basic representation of the individual x is enriched by the vector s, containing a finite number of parameters, which controls the genetic operations transforming the individual x to the next genetic epoch (see Section 5.3.1). Because the vectors s may be different for two individuals of the identical genotype, the set-like representation of P_t delivered by the formula 2.16 has to be used in this case.

3.4 Selection

The first operation that starts to transform the population P_t in the tth epoch according to the scheme of stochastic search 2.2 is *selection operation.* We briefly describe four kinds of selection: proportional (roulette) selection, tournament selection, elitist selection and rank selection. Selection is mainly a random operation (except the pure elitist one) and may be modeled as multiple sampling from the population P_t, although its implementation may take quite different forms.

Roughly speaking, the sampling probability distribution of selection sel_f will be an element of the space of probabilistic measures on P_t

$$sel_f \in \mathcal{M}(P_t). \tag{3.24}$$

The lower index f informs us that it always depends on the fitness function f in some way. The detailed form of sel_f depends on the population cardinality and representation.

We will present three forms of the selection sampling measure denoted by sel^i_f, $i = 1, 2, 3$. In future considerations we will use the same generic description sel_f for all representations. The detailed meaning of this symbol will depend on the particular context.

In the case of the finite population $\#P_t = \mu < +\infty$ we may use the set-like representation (see Remark 2.15)

$$\Big\langle \underbrace{z_1, z_1, \dots, z_1}_{\eta(z_1)\text{ times}}, \underbrace{z_2, z_2, \dots, z_2}_{\eta(z_2)\text{ times}}, \dots, \underbrace{z_\chi, z_\chi, \dots, z_\chi}_{\eta(z_\chi)\text{ times}} \Big\rangle \tag{3.25}$$

where the set $\text{supp}(\eta_t) = \{z_1, \dots, z_\chi\} \subset U,\ \chi \leq \mu < +\infty$. The finite distribution

$$sel^1_f = \{p_1, \dots, p_\mu\} \tag{3.26}$$

may distinguish among two individuals with the same genotype, but this possibility is rarely applied in practise. The probability p_i is assigned to the sampling of the ith element from the list 3.25. Note that the order of elements in 3.25 is crucial in this case.

In this same case $\#P_t = \mu < +\infty$, if the distinction among clones of the same genotype is not necessary, we may process selection as multiple sampling from the set $\text{supp}(\eta_t)$. Selection sampling measure will also be finite in this case

$$sel^2_f = \{p_1, \dots, p_\chi\} \in \mathcal{M}(\text{supp}(\eta_t)) \tag{3.27}$$

where p_i denotes the probability of sampling the genotype $z_i \in \text{supp}(\eta_t)$. The probability of selecting a single individual from P_t with a genotype $z_i \in \text{supp}(\eta_t)$ equals

$$\frac{p_i}{\eta_t(z_i)} \tag{3.28}$$

and is identical for all $\eta_t(z_i)$ such individuals of the same genotype z_i.

Neither of the above representations of sel_f^1, sel_f^2 depend on the cardinality of genetic universum U.

If now $\#U = r < +\infty$ we may easily extend the measure given by the formula 3.27 to the discrete measure defined on the whole genetic space U by the formula

$$sel_f^3 = \{p_1, \ldots, p_r\} \in \mathcal{M}(U);$$

$$sel_f^3(\{x\}) = \begin{cases} sel_f^2(\{x\}) & \text{for } x \in \text{supp}(\eta_t) \\ 0 & \text{otherwise} \end{cases} \tag{3.29}$$

Selection may be understood formally as multiple sampling from U with the sampling measure 3.29 in this case.

If now $\#P_t = +\infty$, but $\#U = r < +\infty$ we may use the population representation postulated by Remark 2.12. Both forms of selection sampling measures sel_f^2, sel_f^3 given by formulas 3.27, 3.29 remain valid except formula 3.28 which evaluates the probability of sampling a single individual, which does not make sense in this case.

3.4.1 Proportional (roulette) selection

Proportional selection may be explained as the multiple sampling from $\text{supp}(\eta_t)$ with a probability distribution $sel_f \in \mathcal{M}(\text{supp}(\eta_t))$ given by the formula

$$sel_f(\{x\}) = \frac{f(x)\,\eta_t(x)}{\sum_{y \in \text{supp}(\eta_t)} f(y)\,\eta_t(y)},\ x \in \text{supp}(\eta_t) \tag{3.30}$$

The denominator $\sum_{y \in \text{supp}(\eta_t)} f(y)\,\eta_t(y)$ in the right hand side of the above formula 3.30 stands for the *total fitness of population* P_t.

The meaning of the attribute "proportional" is immediately seen from formula 3.30. The probability of sampling $x \in \text{supp}(\eta_t)$ is proportional to its impact on the total population fitness expressed by the product of its fitness $f(x)$ and the number of individuals $\eta_t(x)$ of the genotype x in the population P_t.

The meaning of the attribute "roulette" comes from the standard implementation which consists of sampling a single point from the range $[0,1] \subset \mathbb{R}$ that is divided into $\#\text{supp}(\eta_t)$ parts according to the distribution 3.30. Location of the sampled point may be identified with the final position of the roulette wheel with the unit perimeter.

Roulette, proportional selection is one of the few cases in which the sampling probability distribution sel_f is available explicitly. It is also easy to see that if $f|\text{supp}(\eta_t) > 0$, then each individual from the population P_t has a chance to pass to the next step of a genetic algorithm.

3.4.2 Tournament selection

Single sampling in this kind of selection is performed in two steps:

1. Selecting in some way the finite sub-multiset (see definition 2.10) from the population P_t. The selected multiset of the cardinality $k \leq \#P_t$, $k < +\infty$ is called *tournament mate.*
2. Selecting one of the best fitted individuals from the tournament mate already undertaken.

Typical size of the tournament mate is taken as $k = 2$ (the couple of individuals that will take part in the tournament) and the usual methods of mate selection is the simple k-time sampling (without returning) according to the uniform probability distribution on the whole P_t, or by the multiple sampling from $\text{supp}(\eta_t)$ according to the distribution 3.30 (see e.g. Goldberg [74]).

The explicit evaluation of the selection sampling probability distribution is rather complicated in this case. Such evaluation is delivered by Vose [193] under some rather restrictive assumptions for the case of binary genotypes (binary encoding).

3.4.3 Elitist selection

The main idea of elitist selection is to ensure the passage of the best fitted individuals from the current population P_t to the next steps of the algorithm. The sampling procedure may be formalized as a two-phase one:

1. We select in a deterministic way the sub-multiset $Elite \subset P_t$ which gathers the best fitted individuals from the current population. $Elite$ is passed to the next epoch population with the probability 1.
2. Other individuals are selected according to other, probabilistic rules, e.g. the proportional selection rule described in Section 3.4.1.

The most popular case of the elite set is the singleton $Elite = \{\hat{z}\}$ where $\hat{z} \in P_t; f(\hat{z}) \geq f(x)\ \forall x \in P_t$. Because of the partially deterministic way of sampling, it is difficult to express the elitist selection in terms of multiple sampling with the prescribed probability distribution.

3.4.4 Rank selection

Rank selection establishes the arbitrary way to designate the sampling probabilities $sel(\{z\})$, $z \in P_t$ by defining their ranks $R(z), z \in P_t$. The set-like population's representation (see formula 3.25) allows the most general rank assignment in the case of finite populations $\#P_t = \mu < +\infty$.

Perhaps the simplest role of rank assignment is based on the arbitrary total order $\succ$ in P_t represented as the list 3.25 of individuals which satisfies the condition:

$$\forall y, z \in P_t \; y \succ z \Rightarrow f(y) \geq f(z) \tag{3.31}$$

The rank assignment is performed in the following steps:

1. $R(z) = R_0$ for the lowest element in P_t with respect to the $\succ$ order ($\forall y \in P_t;\; y \neq z \; y \succ z$).
2. $R(x) = R(y) + 1$ if y is the immediate predecessor of x in P_t with respect to the $\succ$ order ($y \succ x$ and $\neg\{\exists w \in P_t;\; y \succ w \succ x\}$).

In this case the rank assignment mapping

$$R : P_t \to [R_0, R_0 + \mu] \cap \mathbb{Z}_+ \tag{3.32}$$

is one-to-one (bijection). Another case of rank assignment that may lead to rank ambiguity can be obtained by replacing the second step of the above algorithm by the following one

1. If y is the immediate predecessor of x in P_t with respect to the $\succ$ order and $f(x) > f(y)$ then $R(x) = R(y) + 1$ and $R(x) = R(y)$ if $f(x) = f(y)$.

We may find various formulas that allow us to compute selection sampling probabilities by the given rank distribution 3.32. Two sample formulas of this type are quoted after the monograph [5].

$$\begin{aligned} sel_f(\{z\}) &= a + k\left(1 - \frac{R(z)}{R_{\max}}\right) \\ sel_f(\{z\}) &= a + k\,(R_{\max} - R(z))^b \end{aligned} \tag{3.33}$$

where $R_{\max} = \max_{z \in P_t}\{R(z)\}$ and the parameters $a, k, b \in \mathbb{R}$ are chosen so that the probability distribution sel_f is well defined

$$\sum_{z \in P_t} sel_f(\{z\}) = 1, \; 0 \leq sel_f(\{z\}) \leq 1 \; \forall z \in P_t. \tag{3.34}$$

3.5 Binary genetic operations

Genetic operations are used to produce genotypes of new individuals from one or more genotypes of individuals, which come from the current populations. Binary genetic operations are especially designed to operate on binary genotypes, being the elements of Ω (see formula 3.5). Their implementations may be formalized as the mappings $\Omega^p \to \Omega, \; p \in \mathbb{N}$. Genetic operations have a stochastic character, so their values are random variables and a more accurate description may be stressed as follows

$$\Omega^p \to \mathcal{M}(\Omega), \; p \in \mathbb{N}. \tag{3.35}$$

Our goal will be to present probability distributions from $\mathcal{M}(\Omega)$ associated with the above variables as well as selected implementation issues. We will present the detailed description of various types of binary mutation and crossover, for which $p = 1, 2$.

3.5.1 Multi-point mutation

Binary multi-point mutation consists of producing the binary code $x' \in \Omega$ from the code $x \in \Omega$ by using the *mutation mask* $i \in \Omega$ according to the formula

$$x' = x \oplus i. \tag{3.36}$$

The point-wise, modulo 2 addition of the mask to the parental code results in the bit inversion in x on positions that coincide with the positions of ones in the mask code i.

The mask $i = (a_0, a_1, \ldots, a_{l-1})$ is sampled independently for the mutation of each particular parental code x and independently of the value of this code. Each bit a_i, $i = 1, \ldots, l-1$ of the mask string is sampled independently from the other a_j, $j \neq i$ during the single mask sampling procedure. The probability of sampling 1 in the single bit sampling is assumed to be constant and equals p_m. Taking the above assumptions into account, the probability ξ_i of sampling the particular mask $i \in \Omega$ can be computed using the classical Bernouli model of independent trials with the binary result (see e.g. Billingsley [28]).

$$\xi_i = (p_m)^{(\mathbf{1},i)}(1 - p_m)^{l-(\mathbf{1},i)} \tag{3.37}$$

where

$$\mathbf{1} = \underbrace{(1, \ldots, 1)}_{l-times}$$

denotes the binary vector of l units and $(\mathbf{1}, i)$ the Euclidean scalar product of binary vectors $\mathbf{1}$ and i.

The multi-point binary mutation then has the single parameter $p_m \in [0, 1]$ called *mutation rate*. Assuming that the parental code $x \in \Omega$ is changed by the binary multi-point mutation then the probability distribution of its new value $mut_x \in \mathcal{M}(\Omega)$ is given by the formula

$$mut_x(\{x'\}) = \sum_{i \in \Omega} \xi_i [x \oplus i = x'] \tag{3.38}$$

where $[\cdot]$ denotes the binary evaluation of the logical expression so that

$$[w] = \begin{cases} 0 & \text{if } w \text{ is true} \\ 1 & \text{if } w \text{ is false} \end{cases} \tag{3.39}$$

3.5.2 Binary crossover

Binary crossover is the operation that produces the binary code $z \in \Omega$ from two parental codes $x, y \in \Omega$ by using the binary string $i \in \Omega$ called *crossover mask*. The code z is sampled from the pair of two "children"

$$\left\{ (y \otimes i) \oplus (\hat{\imath} \otimes x), \, (x \otimes i) \oplus (\hat{\imath} \otimes y) \right\} \tag{3.40}$$

with the uniform probability distribution $\{\frac{1}{2}, \frac{1}{2}\}$. The binary vector $\hat{i} = \mathbf{1} \oplus i \in \Omega$ stands for the inversion of the mask i. The action of the mask in formula 3.40 results in bit exchanging among parental codes on positions on which appears 1 in the mask code.

The crossover mask i is sampled from Ω according to the probability distribution $(\zeta_0, \zeta_1, \dots, \zeta_{r-1}) \in \mathcal{M}(\Omega)$ given by the formula

$$\zeta_i = \begin{cases} p_c \, type_i & i > 0 \\ 1 - p_c + p_c \, type_0 & i = 0 \end{cases} \tag{3.41}$$

where the vector $type = (type_0, type_1, \dots, type_{r-1}); \sum_{i \in \Omega}, type_i \geq 0 \, \forall i \in \Omega$ constitutes another probability distribution called *crossover type* in the space $\mathcal{M}(\Omega)$. The number $p_c \in [0, 1]$ will be called the *crossover rate*.

For the most classical one-point crossover introduced by Holland [85] the crossover type takes the form

$$type_j = \begin{cases} \dfrac{1}{l-1} & \text{if } \exists k \in (0, l) \cap \mathbb{Z};\ j = 2^k - 1 \\ 0 & \text{otherwise} \end{cases} \tag{3.42}$$

It is easy to see that the above distribution assigns the non-zero, equal probability $\frac{1}{l-1}$ to each string of the form

$$j = 2^k - 1 = \left(\underbrace{0, \dots, 0}_{l-k-\text{times}}, \underbrace{1, \dots, 1}_{k-\text{times}} \right), \quad k = 1, 2, \dots, l-1. \tag{3.43}$$

Other strings are postponed by the zero probability. Coupling both formulas 3.41 and 3.42 we obtain the mask probability distribution for one-point crossover

$$\zeta_i = \begin{cases} 1 - p_c & i = 0 \\ \dfrac{p_c}{l-1} & i = 2^k - 1,\ k = 1, 2, \dots, l-1 \\ 0 & \text{otherwise} \end{cases} \tag{3.44}$$

The probability $1 - p_c$ is assigned to only one mask $i = 0$, which performs trivial crossover (simply no crossover, the resulting string z is sampled from the parents x, y only). Note that the mask $i = \mathbf{1}$ that also trivially transforms parents to children has a zero sampling probability. All masks of the type 3.43 have the same, non-zero probability $\frac{p_c}{(l-1)}$ and refer to the non-trivial crossover operation. This operation results in tearing the parent strings between the k and $k+1$ positions and then replacing the obtained parts among them.

The formal description presented above was introduced by Vose (see e.g. [193]) and allows us to flexibly handle other types of binary crossover met in practice. One important example is the *uniform crossover* for which the *type* distribution is given by

$$type_i = 2^{-l}, \ \forall i \in \Omega \tag{3.45}$$

The trivial crossover masks $i = 0, i = \mathbf{1}$ now have the total sampling probability $1 - \left(1 - \frac{1}{2^{l-1}}\right) p_c$ while each other mask $i \in \Omega; i \neq 0, i \neq \mathbf{1}$ has the equal probability $\frac{p_c}{2^l}$.

The binary crossover operation then has two parameters: the crossover rate $p_c \in [0,1]$ which controls the overall rate of crossover intensity and $type \in \mathcal{M}(\Omega)$ which specifies the kind of bit exchanging operation among parents.

The probability distribution $cross_{x,y} \in \mathcal{M}(\Omega)$ that determines the stochastic result of crossing strings $x, y \in \Omega$ is given by the formula

$$cross_{x,y}(\{z\}) = \sum_{k \in \Omega} \frac{\zeta_k + \zeta_{\hat{k}}}{2} [(x \otimes k) \oplus (\hat{k} \otimes y) = z] \tag{3.46}$$

3.5.3 Features of binary genetic operations, mixing

We will concentrate in this section on the crucial features of binary genetic operations, which allow us to formulate rigorous mathematical models of genetic algorithms that handle binary encoding and may facilitate their implementation. All results are taken from or strictly based on Vose's monograph [193].

Definition 3.9. *The binary mutation described in Section 3.5.1 will be called independent if the probability ξ_j of sampling the arbitrary mutation mask $j \in \Omega$ satisfies the condition*

$$\xi_j = \sum_{k \otimes i = 0} \xi_{i \oplus j} \sum_{\hat{k} \otimes i = 0} \xi_{i \oplus j} \ \ \forall k \in \Omega \tag{3.47}$$

□

Theorem 3.10. *(see Vose [193], Theorem 4.1) If the mutation mask is sampled with the probability distribution given by formula 3.37, then mutation is independent.* □

The mutation independence can affect the coupling of both crossover and mutation in the following way.

Theorem 3.11. *(see Vose [193], Theorem 4.2) If the binary mutation is independent, then, for the arbitrary binary crossover the two following events have the same probability:*

- *Obtain the string $z \in \Omega$ by crossing x' and y', where x', y' resulting from the mutation of codes x, y respectively.*
- *Obtain the string $z \in \Omega$ mutating the result of crossover of strings x, y.* □

The next two theorems show the common probabilistic effect of mutation and crossover composition without any additional assumptions concerning mutation.

Theorem 3.12. *(see Vose [193], Theorem 4.3) If we mutate two strings and next cross their offspring, then the common probability distribution $m_{x,y} \in \mathcal{M}(\Omega)$ of obtaining a new string z from the parental strings x, y is given by the formula*

$$m_{x,y}(\{z\}) = \sum_{i,j,k \in \Omega} \xi_i \, \xi_j \frac{\zeta_k + \zeta_{\hat{k}}}{2} \left[(x \oplus i) \otimes k \oplus \hat{k} \otimes (y \oplus j) = z \right] \tag{3.48}$$

If mutation is applied to x, y which are already crossed, then

$$m_{x,y}(\{z\}) = \sum_{j,k \in \Omega} \xi_j \frac{\zeta_k + \zeta_{\hat{k}}}{2} \left[x \otimes k \oplus \hat{k} \otimes y \oplus j = z \right] \tag{3.49}$$

□

Theorem 3.13. *(see Vose [193], Theorem 4.3) Both probability distributions, given by formulas 3.48, 3.49 satisfy the conditions*

$$m_{x,y}(\{z\}) = m_{y,x}(\{z\}) = m_{x \oplus z, y \oplus z}(\{\mathbf{0}\}) \tag{3.50}$$

where $\mathbf{0} \in \Omega$ stands for the binary vector of l zeros. □

Remark 3.14. If the binary mutation is independent, then both probability distributions given by formulas 3.48, 3.49 are equivalent which is the result of Theorem 3.12. It holds, in particular, for multi-point mutation with the mask sampling according to formula 3.37 (see Theorem 3.10) and the binary crossover, for the arbitrary operation parameters p_m, p_c and *type*. □

Definition 3.15. *If the mutation is independent, then we may define the operation called* mixing *as a composition of mutation and crossover. The order of component operation does not affect its probabilistic result, so the probability distribution of mixing may be computed using both formulas 3.48, 3.49.* □

3.6 Definition of the Simple Genetic Algorithm (SGA)

The Simple Genetic Algorithm (SGA) is a method of transforming a binary represented population P_t to the next epoch population P_{t+1}. Both samples are multisets of binary strings (genotypes) from the binary genetic universum Ω (see formula 3.5 in Section 3.1.1).

We have to fix five parameters of SGA: the population size $\mu \in \mathbb{N}$, the fitness function f, mutation and crossover rates $p_m, p_c \in [0, 1]$ and the crossover type $type \in \mathcal{M}(\Omega)$, which is the probabilistic vector of the length r. All parameters remain unchanged during the whole computation.

The algorithm consists of running in the finite loop, three random operations described in items 2–4 until the condition contained in item 5 is satisfied.

1. Create an empty population P_{t+1}.
2. Select two individuals x, y from the current population P_t by multiple sampling according to the probability distribution 3.30 (proportional, roulette selection).
3. Produce the binary code $z \in \Omega$ from selected $x, y \in \Omega$ using mixing with the probability distribution 3.48 (or 3.49).
4. Put z to the next epoch population P_{t+1}.
5. If P_{t+1} contains less than μ individuals, go to 2.

SGA is one of a few instances of genetic computations for which the probability distribution of sampling the next epoch population can be delivered explicitly. This problem will be discussed in Chapter 4.

3.7 Phenotypic genetic operations

Phenotypic genetic operations are used in genetic computation models in which the genetic universum U is simply the admissible set $\mathcal{D}$ which is the subset of the finite dimensional space V usually identified with $\mathbb{R}^N$ (see Sections 2.1, 3.1.3). Basic versions of most phenotypic operations have the form of simple geometric formulas, which utilize the multidimensional normal probability distribution (see e.g. [28]).

Binary encoded genetic algorithms search only in $\mathcal{D}$ by checking the discrete subset $\mathcal{D}_r \subset \mathcal{D}$ of phenotypes. They meet the constraints imposed by the global optimization problem, which restrict its solution to the set $\mathcal{D}$.

For genetic algorithms with the phenotypic encoding (frequently called *evolutionary algorithms*) the problem of restricted search in the bounded domain $\mathcal{D}$ is usually non-trivial (see e.g. Arabas [5], Bäck, Fogel, Michalewicz [15]). Such a problem may be solved by modifying the objective function by adding a penalty or by modification of genetic operations in such a way that they will turn back only the individuals representing the admissible points in $\mathcal{D}$. Some remarks considering constrained genetic search by phenotypic encoding will be placed in Section 3.7.3. Readers interested in this matter are referred to books [5, 15, 11, 110].

3.7.1 Phenotypic mutation

Mutation stands for the basic, most important genetic operation in evolutionary algorithms. One of the simplest and most frequently used kinds of phenotypic mutation is called *normal phenotypic mutation* (see e.g. Schwefel [163]). It creates the offspring individual x' from the parental one $x \in U = \mathcal{D}$ by the formula

$$x' = x + \mathcal{N}(0, \mathbf{C}) \tag{3.51}$$

where $\mathcal{N}(0, \mathbf{C})$ is the realization (result of sampling) of the N dimensional random variable that has the multidimensional, normal distribution with mean $0 \in V$ and the $N \times N$ dimensional covariance matrix $\mathbf{C}$ that constitutes the parameters of this operation. Please note that because the normal perturbation $\mathcal{N}(0, \mathbf{C})$ may take an arbitrary value in V with a positive probability, then also x' may lie in the whole V regardless of whether the parent x is in $\mathcal{D}$ or not.

Many other mutation models are based on the formula 3.51 of genotype modification, changing only the probability distribution of perturbation which is here $\mathcal{N}(0, \mathbf{C})$. A review and detailed discussion of such solutions may be found in broadly cited monographs [5, 15, 11, 110].

One new and interesting way of improving perturbation distributions in many dimensions was shown by Obuchowicz [119] who tries to eliminate the "surrounding effect" that appears by normal mutation. This effect consists of concentrating offspring x' at a distance from x that approximately equals the mean eigenvalue of $\mathbf{C}$.

A completely different type of mutation that may be applied in evolutionary algorithms falls into the group of Lamarcean operations, using a local optimization method for making individual perturbation. Such an approach will be described and roughly discussed in Section 5.3.4.

3.7.2 Phenotypic crossover

The simple phenotypic crossover rule, frequently called *arithmetic crossover*, may be given after Wright [203].

$$x' = x^1 + \mathcal{U}[0,1]\left(x^2 - x^1\right) \tag{3.52}$$

where $x^1 = \left(x_1^1, \ldots, x_N^1\right), x^2 = \left(x_1^2, \ldots, x_N^2\right) \in \mathcal{D}$ are parental individuals, $\mathcal{U}[0,1]$ is the realization (result of sampling) of the one-dimensional random variable with a uniform distribution over the interval $[0, 1]$. Using the above formula we obtain the child x' located inside the N-dimensional interval $[x^1, x^2]$. This formula may ensure the coherency of crossover ($x' \in \mathcal{D}\ \forall x \in \mathcal{D}$) only if $\mathcal{D}$ is strictly convex.

Arabas [5] delivers an alternative formula to 3.52

$$x_i' = x_i^1 + \mathcal{U}[0,1]\left(x_i^2 - x_i^1\right),\ i = 1, \ldots, N \tag{3.53}$$

in which we use the separate realization of the random variable $\mathcal{U}[0,1]$ in each direction. The resulting child individual x' will be located in the N-dimensional segment spanned by the parents x^1, x^2 so it is not necessarily included in $\mathcal{D}$ even if the admissible set is convex.

Another possibility proposed by Michalewicz [111] uses geometric means for the parental coordinates in order to obtain the child coordinates

$$x_i' = \sqrt{x_i^1\, x_i^2},\ i = 1, \ldots, N \tag{3.54}$$

Note that this operation is deterministic with respect to parents already selected. The above operation is called *geometric crossover.*

Both arithmetic and geometrical crossover may be easily extended to the operations that can cross $k > 2$ parents $x^1, x^2, \dots, x^k \in \mathcal{D}$ (see e.g. [15]). In particular, the barycentric combination of parents

$$x' = \sum_{i=1}^{k} \alpha_i x^i, \; \alpha_i \geq 0, i = 1, \dots, k, \; \sum_{i=1}^{k} \alpha_i = 1 \tag{3.55}$$

extends the arithmetic crossover 3.52, while

$$x'_i = \prod_{j=1}^{k} \left(x_i^j\right)^{\alpha_j}, \; i = 1, \dots, N \tag{3.56}$$

extends the geometric crossover 3.54. The generic coefficients $\alpha_1, \dots, \alpha_k$ may be set in a deterministic or random way.

Another possibility to define multi-parent crossover is the *simplex crossover* introduced by Renders and Bersini [135]. Two parents, the worst fitted $\check{x}$ and the best fitted $\hat{x}$ among the parent mate $x^1, x^2, \dots, x^k \in \mathcal{D}$, were selected. Next, the centroid c of the parent mate without $\check{x}$ was computed. The offspring is the reflection of $\hat{x}$ with respect to the centroid

$$x' = c + (c - \hat{x}) \tag{3.57}$$

The fitness-based multi-parent crossover was also utilized by Eiben et all [63].

3.7.3 Phenotypic operations in constrained domains

As we mentioned at the beginning of Section 3.7 phenotypic genetic operations do not generally exhibit the coherency condition $\Pr\{x' \in \mathcal{D}\} = 1, \; \forall x \in \mathcal{D}$. The following ways of handling the constraints imposed by the global optimization problem are possible:

1. Standard modification of a constrained global optimization problem to an unconstrained one by transformation of coordinates or introducing the penalty function. This approach is described in many monographs e.g. [66, 5].
2. The special kind of encoding in which the assumed genetic operations will be coherent. Such a condition is always satisfied if $U = \mathbb{R}^N$. This method mainly falls into the case of coordinate transformation mentioned previously.
3. Special kinds of genetic operations that do not exceed the admissible domain $\mathcal{D}$. The examples of such operations (boundary mutation) may be found in books [110, 5].

4. Additional genetic operations called *reparing operations* which are functions of the type $R : V \to \mathcal{D}$. They affect each offspring x' obtained from the parent(s) in the following way

$$R(x') = \begin{cases} x' & \text{if } x' \in \mathcal{D} \\ x'' \in \mathcal{D} & \text{if } x' \notin \mathcal{D} \end{cases} \tag{3.58}$$

   where x'' is the "repaired", admissible individual. The action of the operation consists mainly of projection of x' on the boundary $\partial\mathcal{D}$ according to the prescribed projection rules. Such a rule results usually in enormous growth of boundary individuals, which may decrease the exploration skill of the evolutionary algorithm under consideration. Another possibility of individual repairing will be suggested in the next item.
5. One possible way to create the repairing operations that does not increase the number of boundary individuals is to utilize the "internal reflection rule". Let us consider the vector $x' - x^1$ that joins the offspring with the first of its parents x^1. If $x' \notin \mathcal{D}$ then there is a $(x')^1 \in \partial\mathcal{D}$ which is the first boundary point in the direction $x' - x^1$. We will compute a new point $(x'')^1$ so that the vector $(x'')^1 - (x')^1$ is the vector that lies in the reflection direction satisfying the condition

$$\left(\frac{(x'')^1 - (x')^1}{\|(x'')^1 - (x')^1\|}, \mathbf{n} \right) = \left(\frac{(x)^1 - x'}{\|(x)^1 - x'\|}, \mathbf{n} \right)$$

   where $\mathbf{n}$ in the internal normal versor of the boundary $\partial\mathcal{D}$ at the point $(x')^1$. Moreover,

$$\|(x'')^1 - (x')^1\| + \|(x)^1 - x'\| = \|x' - x\| .$$

   If $(x'')^1 \in \mathcal{D}$ then $x'' = (x'')^1$. If not, then the procedure is repeated until the consecutive j-th point satisfies $(x'')^j \in \mathcal{D}$. The determination of the reflection points $(x')^i \in \partial\mathcal{D}$, $i = 2, \ldots, j-1$ and the normal vectors at these points are necessary. If for any $(x')^i$ the normal versor is not unambiguously determined, then we set the reflection direction as opposite to the falling direction. The repairing procedure should act properly in the case of admissible domains with the Lipschitz boundary $\partial\mathcal{D}$ (see Section 2.1).

A broad review of the methods of the genetic constrained optimization may be found in the monographs [5, 15, 11, 110].

3.8 Schemes for creating a new generation

Genetic algorithms may offer a more general routine for making the next epoch population P_{t+1} than SGA. We may distinguish two phases of such a

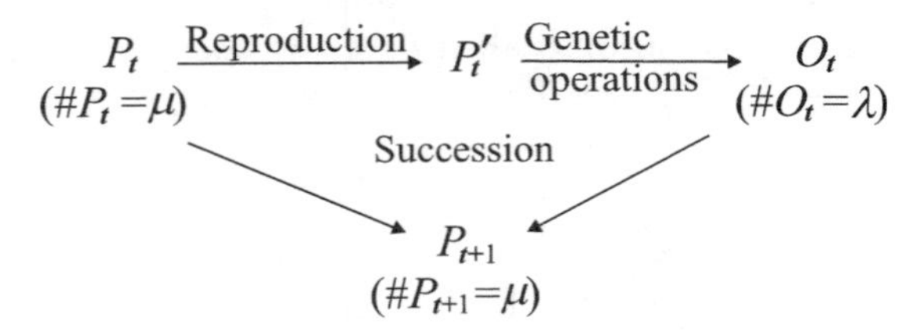

Fig. 3.2. Producing new population by reproduction and succession

routine: reproduction and succession (see [5]). Relations among phases were shown in Figure 3.2.
Reproduction consists of creating the intermediate sample P_t' by multiple sampling from the current population P_t. The individuals that belong to P_t' will be called *parental individuals*. Reproduction is performed by using one of the selection techniques described in Section 3.4. The cardinality of the intermediate sample $\#P_t'$ is one of the important reproduction parameters.

Genetic operations process all parental individuals from P_t' to offspring individuals that form the new multiset O_t called simply *offspring*.

Succession makes the next epoch population P_{t+1} by sampling only from offspring O_t in the case of so-called (μ, λ)-type genetic algorithms, or from the multiset $P_t \cup O_t$ in the case of so-called $(\mu + \lambda)$-type algorithms. According to the widely-accepted Schwefel notation discussed in the next section, the parameters μ and λ stand for the cardinality of the current and next epoch populations $\mu = \#P_t = \#P_{t+1}$ and the offspring cardinality $\lambda = \#O_t$ respectively. Such a rule may be broken in the case of algorithms with a variable life time of individuals, as mentioned in Section 5.3.2.

Succession is performed mainly by the proper selection technique, but usually a different type than in the case of reproduction. The most popular kind of succession is the pure elitist one in which $\mu = \#Elite$, then we select μ best individuals from $P_t \cup O_t$ or only from O_t with the probability 1.

3.9 μ, λ – taxonomy of single- and multi-deme strategies

We try to comprehend the main idea of the useful notation introduced by Schwefel [163] and then extended by Rechenberger [133] and Rudolph [145]. This notation allows us to distinguish the various types of succession that may happen in the classical scheme of stochastic searches, in which a single sample called population is processed, as well as in the case in which the structure of *demes*, sometimes called subpopulations, is processed.

$(\mu + \lambda)$ Algorithms of such types produce λ offspring individuals then samples μ individuals from $P_t \cup O_t$ in order to createP_{t+1} in each epoch. The notation probably comes from the cardinality of the sampling domain $\mu + \lambda = \#P_t \cup O_t$. The particularly important cases are: $(1+1)$ genetic random walk with only mutation operation, the $(\mu + 1)$ algorithm that reproduces only one individual in each epoch.

(μ, λ) Algorithms of such types produce λ offspring individuals then samples μ individuals from O_t in order to create P_{t+1} in each epoch. We usually assume that $\lambda \geq \mu$ in order to prevent the genetic material degeneration. The case $\mu = \lambda = 1$ is utilized only by the classical random walk search.

$(\mu \overset{+}{,} \lambda)$ This notation comprehends both groups of algorithms mentioned before (exactly as their union). The kind of succession is arbitrary, only the population size μ and the offspring number λ is crucial.

(μ, κ, λ) This group gathers strategies of $(\mu \overset{+}{,} \lambda)$ type for which selection is performed by comparison of the individual life time with its maximum value κ. Such strategies will be described in Section 5.3.2.

$[\mu' \overset{+}{,} \lambda' (\mu \overset{+}{,} \lambda)^{\gamma}]^{\gamma'}$ This is a group of multi-deme, nested genetic algorithms. The random sample is composed of μ' demes, each of the cardinality μ. They were used to produce λ' new demes, each of the cardinality μ. For these new demes, the $(\mu \overset{+}{,} \lambda)$ genetic algorithm is processed by γ epochs. The overall multi-deme scheme is processed γ' epochs. Multi-deme genetic algorithms will be discussed in Section 5.4.

According to the above taxonomy and notation, the Simple Genetic Algorithm SGA is of the (μ, λ) type. Moreover, $\mu = \lambda$ and reproduction is performed by using proportional selection. Succession is performed by the elitist selection for which $\#Elite = \mu$ (the whole offspring O_t passes to the next population P_{t+1}).

4

Asymptotic behavior of the artificial genetic systems

This chapter discusses the possibility of the formal analysis of stochastic global genetic searches that utilizes genetic-like mechanisms to obtain the consecutive population from the current one. Their asymptotic behavior is interesting, as is usual in the case of iterative algorithms. Most definitions have been introduced and features proved for genetic algorithms that can be modeled as the Markov chain with the space of states that represents all possible populations. Selected results of the Markov approach to the Simple Genetic Algorithm and special instances of evolutionary algorithms obtained in groups led by Michael Vose and Günter Rudolph were reported in the first, dominating Section 4.1. The study of genetic algorithm sampling measures will be proposed as the new tool for the analysis in cases of continuous global optimization problems (see Section 4.1.2). The Markov approach to very small population dynamics was reported in Section 4.2. Moreover, the classical schemata approach to the analysis of the single-step transition of the Simple Genetic Algorithm was revisited in Section 4.3.

4.1 Markov theory of genetic algorithms

The application of the Markov theory of stochastic processes with discrete time in the analysis of genetic algorithms has to be preceded by the definition of the space of states that can fully characterize the progress of genetic computation. This space will be constructed on the basis of the genetic universum U that may contain only a finite or infinite, continuous number of codes. One important case is the binary genetic universum $\Omega, \#\Omega = r < +\infty$ (see formula 3.5).

Definition 4.1. *The space of states $\mathcal{E}$ of the genetic algorithm will be the set to which all populations (or their unambiguous representations) that may be produced by this algorithm belong.* □

R. Schaefer: *Foundation of Global Genetic Optimization*, Studies in Computational Intelligence (SCI) **74**, 55–113 (2007)
www.springerlink.com

If the individual is fully characterized by its genotype, which is true for almost all the algorithms described in Chapter 3, then the space of states satisfies the inclusion

$$\mathcal{E} \subset U^{\mu}/eqp, \; 0 < \mu < +\infty \tag{4.1}$$

where μ denotes the finite, common cardinality of all the populations created by the genetic algorithm, and *eqp* the relationship defined by the formula 2.14, in which $\mathcal{Z}$ is substituted by U.

If the genetic universum is finite $\#U = r < +\infty$ (which in particular holds for the binary universum Ω), then we may identify it with the finite set of integer indices

$$U \sim \{0, 1, \ldots, r-1\} \tag{4.2}$$

In this case, each population $P \in \mathcal{E}$ may be identified with its frequency vector

$$P \sim (x_0, x_1, \ldots, x_{r-1}); \; x_i \in [0,1], \; \sum_{i=0}^{r-1} x_i = 1, \; x_i = \frac{\eta(i)}{\mu} \tag{4.3}$$

where η is the occurrence function of the population, namely $P = (U, \eta)$ according to the multiset definition 2.8.

The above relation shows that, assuming finite genetic universum, the space of states $\mathcal{E}$ may be identified with the finite set X_{μ} which is the subset of the $(r-1)$-dimensional simplex Λ^{r-1} in $\mathbb{R}^r$.

$$\begin{aligned} \mathcal{E} \sim X_{\mu} \subset \Lambda^{r-1} = \\ = \left\{ x = (x_0, x_1, \ldots, x_{r-1}); 0 \leq x_i \leq 0, i = 0, \ldots, r-1, \sum_{i=0}^{r-1} x_i = 1 \right\} \end{aligned} \tag{4.4}$$

Remark 4.2. (see Vose [193]) If $\mu < +\infty$ and $\#U = r < +\infty$ (the case of finite population and finite genetic universum) then the number of possible populations is also finite and equals

$$n = \#X_{\mu} = \binom{r + \mu - 1}{\mu - 1} < +\infty$$

where $\binom{\cdot}{\cdot}$ denotes Newton's binomial operator. □

If the population cardinality μ grows, then the number of possible states $n = \#X_{\mu}$ will increase even if the number of genetic codes remains unchanged $\#U = r = constant$. Because the elements of X_{μ} are evenly distributed in Λ^{r-1}, then the following remark may be drawn.

Remark 4.3. (see Nix and Vose [116], Vose [191, 193])

$$\overline{\lim_{\mu \to +\infty} X_\mu} = \Lambda^{r-1}$$

□

The above remark allows us to study the abstract genetic algorithms that utilize the finite genetic universum and work with infinite, yet countable populations ($\#U = r < +\infty$, $\#P = \#\mathbb{Z}$). The space of states of such algorithms $\mathcal{E} = \Lambda_{r-1}$ is infinite, but bounded and compact in $\mathbb{R}^r$.

Let us turn our attention to several characteristic states of such algorithms:

Remark 4.4.

- Vertices $x^{(j)} = (0, \ldots, 0, \overset{(j)}{1}, 0, \ldots, 0)$, $j = 1, \ldots, r$ of the simplex Λ^{r-1} represent populations (P, η) so that $\eta(j) > 0, \eta(k) = 0, k \neq j$. We sometimes call such populations *monochromatic* ones.
- The center $x^{(S)} = \left(\frac{1}{r}, \ldots, \frac{1}{r}\right)$ of the simplex Λ^{r-1} represents the population (P, η) in which all genotypes are uniformly represented ($\eta(i) = \eta(j) > 0, \forall i, j \in U$). □

4.1.1 Markov chains in genetic algorithm asymptotic analysis

Each genetic algorithm may be interpreted as a system that transforms the population $P_t \in \mathcal{E}$ into another one $P_{t+1} \in \mathcal{E}$ in each genetic epoch $t = 0, 1, 2, \ldots$ then produces the sequence $\{P_t\}, t = 0, 1, 2, \ldots$. Because this transformation has a stochastic character, the populations $\{P_t\}, t = 0, 1, 2, \ldots$ may be handled as a family of random variables defined on the common probabilistic space $(\Omega_P, \Sigma, \Pr)$ and take their values in the space of states $\mathcal{E}$. The random sequence is associated with the family of probability distributions, $\{\pi^t\}, t = 0, 1, 2, \ldots$ which are measures from the space $\mathcal{M}(\mathcal{E})$ of probabilistic measures defined over the space of states $\mathcal{E}$. The action of the genetic algorithm in the single genetic epoch t may be explained as one-time sampling from $\mathcal{E}$ according to the probability distribution $\pi^t \in \mathcal{M}(\mathcal{E})$. More precisely,

$$\forall t = 0, 1, 2, \ldots, \forall A \subset \mathcal{E},\ A \,\text{measurable in}\ \mathcal{E}\ \ \Pr\{P_t \in A\} = \pi^t(A). \tag{4.5}$$

The sequence of populations produced by a genetic algorithm may be modeled as a stochastic process with the discrete time $t = 0, 1, 2, \ldots$ and the space of states $\mathcal{E}$.

Remark 4.5. The passage from the population P_t to the next epoch population P_{t+1} is implemented by multiple use of selection and proper genetic operations rather than single-time sampling according to the probability distribution π^t. The effective determination of π^t is rarely possible. Even if it is possible, this way is computationally much more expensive. □

Remark 4.6. The probability distribution $\pi^t \in \mathcal{M}(\mathcal{E})$ can generally depend on all populations $P_0, P_1, P_2, \ldots, P_{t-1}$ and on the vector $u(t)$ of parameters that control genetic operations in the epoch t. This general case will appear in the case of adaptive strategies described in Chapter 5. □

For classical genetic algorithms described in Chapter 3, genetic and selection operations have no memory of previous epochs, i.e. do not utilize any information from previous populations $P_0, \ldots, P_{t-1}$ by producing P_{t+1}. This feature allows the sequence $\{P_t\}, t = 0, 1, 2, \ldots$ to satisfy the Markov condition

$$\Pr\{P_{t+1} \in A | P_0, P_1, P_2, \ldots, P_t\} = \Pr\{P_{t+1} \in A | P_t\}, \ t = 0, 1, 2, \ldots \quad (4.6)$$

where A is an arbitrary measurable set in $\mathcal{E}$.

If the parameters of selection and all genetic operations are constant (do not change with respect to the epoch number), then the following shift condition will be additionally satisfied.

$$\Pr\{P_{t+k} \in A | P_{s+k}\} = \Pr\{P_t \in A | P_s\}, \ \forall k, s \in \mathbb{N} \cup \{0\}, \forall t \in \mathbb{N}, t > s \quad (4.7)$$

The above considerations allow us to observe:

Remark 4.7. If the parameters of the genetic operations (mutation and crossover operation parameters) and the selection parameters are constant, then the classical genetic algorithms described in Sections 3.5, 3.6 and 3.7 can be modeled by using the uniform Markov chain with the space of states $\mathcal{E}$ and the Markov transition function

$$\tau : \mathcal{E} \rightarrow \mathcal{M}(\mathcal{E}); \tau(P_t) = \pi^{t+1}, t = 0, 1, 2, \ldots$$

and the initial distribution $\pi^0 \in \mathcal{M}(\mathcal{E})$ ($\Pr\{P_0 \in A\} = \pi^0(A), A \subset \mathcal{E}$, A is the measurable set, see Billingsley [28]). Moreover, both conditions 4.6, 4.7 hold. □

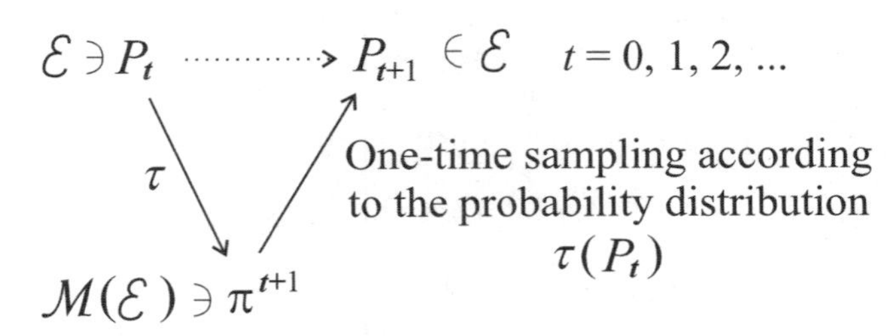

Fig. 4.1. Markov sampling scheme for the classical genetic algorithms.

The scheme of state evolution for classical genetic algorithms is shown on the diagram (see Figure 4.1). The dotted arrow in this figure represents the passage between two consecutive populations by implementing selection and stochastic genetic operations.

Remark 4.8. The mapping $\tau^{(t)} : \mathcal{E} \rightarrow \mathcal{M}(\mathcal{E})$ so that for each measurable set $A \subset \mathcal{E}$ and $P \in \mathcal{E}$

$$\tau^{(t)}(P)(A) = \begin{cases} \tau(P)(A) & t = 1 \\ \int_{\mathcal{E}} \tau^{(t-1)}(y)(A)\tau(P)(dy) & t > 1 \end{cases}$$

will be called the iterate of degree $t = 0, 1, 2, \ldots$ of the Markov transition function τ. □

Remark 4.9. If the genetic algorithm can be modeled by the uniform Markov chain then for each measurable set $A \subset \mathcal{E}$

$$\Pr\{P_t \in A\} = \begin{cases} \pi^0(A) & t = 0 \\ \int_{\mathcal{E}} \tau^{(t)}(y)(A)\pi^0(dy) & t > 1 \end{cases}$$

where $\pi^0 \in \mathcal{M}(\mathcal{E})$ is the initial probability distribution. □

Remark 4.10. Let the conditions of the genetic algorithm assumed in remark 4.7 hold. Moreover, if the transition between consecutive populations $P_t, P_{t+1}, t = 0, 1, 2, \ldots$ is performed by the sequential application of $k \in \mathbb{N}$ selection and genetic operations then the Markov transition function τ may be composed from the transition functions $\tau_1, \ldots, \tau_k$ associated with particular operators

$$\tau(P)(A) = (\tau_1 \circ, \ldots, \circ \tau_k)(P)(A) =$$

$$= \left(\prod_{j=1}^{k-2} \int_{\mathcal{E}} \tau_j(y_j)(dy_j + 1) \right) \int_{\mathcal{E}} \tau_{k-1}(y_{k-1})(dy_k)\tau_k(y_k)(A)$$

where $y_1 = P$, and $y_j, j = 2, \ldots, k$ are the intermediate populations produced by consecutive operations. □

If the space of states of the genetic algorithm is finite $\#\mathcal{E} = n < +\infty$ then the Markov transition function τ may be described by the finite number of transition probabilities which are the entries of the matrix $\mathbf{Q}$ of dimension $n \times n$

$$(\mathbf{Q})_{P,P'} = \tau(P)(\{P'\}), \; P, P' \in \mathcal{E} \tag{4.8}$$

The probability distributions $\{\pi^t\}, t = 0, 1, 2, \ldots$ are also discrete and finite in this case (probability vectors of the dimension $n = \#\mathcal{E}$). The well known discrete Markov transition rule holds

$$\tau^{(t)}(P)(\{P'\}) = (\mathbf{Q}^t)_{P,P'}, \; t = 0, 1, 2, \ldots \tag{4.9}$$

where $\mathbf{Q}^t$ denotes the t-th power of the matrix $\mathbf{Q}$.

If the genetic algorithm with the finite number of states produces a new population by sequential application of $k \in \mathbb{N}$ operations then, similarly as in Remark 4.10 the probability transition matrix $\mathbf{Q}$ can be computed as the product

$$\mathbf{Q} = \prod_{i=1}^{k} \mathbf{Q}^{(i)} \tag{4.10}$$

where the matrices $\mathbf{Q}^{(i)}, i = 1, \ldots, k$ are associated with the consecutive operations.

If the genetic algorithm utilizes the finite genetic universum $\#U = r < +\infty$ and processes finite populations $\#P = \mu < +\infty$ then entries of the probability transition matrix $\mathbf{Q}$ may be indexed by pairs of states from the set $X_\mu \subset \Lambda^{r-1}$, so

$$\mathbf{Q} = \{(\mathbf{Q})_{x,x'}\}\ ;\ x, x' \in X_\mu \subset \Lambda^{r-1} \tag{4.11}$$

The formal approaches presented in this section are cited mainly from the series of papers published by Michael Vose and his collaborators [191, 193, 116], Rudolph [143, 144], Beyer and Rudolph [27], Grygiel [80, 82] and mathematical monographs Feller [65], Chow and Teicher [49], Billingsley [28].

The above results constitute the basis for further research in the following directions:

- Determining the Markov transition function τ or the transition probability matrix $\mathbf{Q}$ for particular genetic operations and algorithms.
- Studying the asymptotic behavior of genetic iterations by studying features of probability transitions in the single step of a genetic algorithm (the passage from one genetic epoch to the consecutive one).
- The particular case of the previous item may consist of studying the ergodicity of the Markov chain which models the genetic algorithm behavior. This feature usually implies the asymptotic correctness in the probabilistic sense or the asymptotic guarantee of success (see definitions 2.16, 2.17) as well as the global convergence (see definition 2.18).
- Finding the mapping called *heuristics* for the particular class of genetic algorithms. This mapping is the transition rule of the idealized instance of the algorithm, which processes the infinite populations. Its features deliver important information about the potential searching ability of this class of algorithms. In particular, the existence and stability of the fixed points of heuristics are of great importance (see Section 4.1.2).
- Studying the existence and features of the invariant measures on the space of states $\mathcal{E}$, i.e. measures which are limits of the sequences $\{\pi^t\}, t \to +\infty$ of measures that determine the population sampling in consecutive genetic epochs (see Remark 4.7).

- Examining the dynamics of sampling measures defined on the search domain $\mathcal{D}$, induced by populations (see formula 3.4). The results of such considerations will be helpful in the justification of finding the central parts of the basins of attractions of global and local extrema of the objective function (see problem $\mathbf{\Pi_4}$, Section 2.1).

4.1.2 Markov theory of the Simple Genetic Algorithm

The Simple Genetic Algorithm (SGA) defined in Section 3.6 works with binary genotypes of the form $(a_0, \ldots, a_{l-1}), a_i \in \{0, 1\}$ where the length of the code $l \in \mathbb{N}$ stands for a fixed parameter for each particular SGA instance. Binary codes constitute the genetic universum Ω so that $\#\Omega = r = 2^l$ (see Section 3.1.1). Such algorithms processing populations of the finite, constant cardinality μ so that $1 < \mu < +\infty$ have the finite space of states $\mathcal{E} = X_\mu = n < +\infty$ whose cardinality may be computed using the formula contained in Remark 4.2. States of SGA may be identified with the population frequency vectors $x = (x_0, \ldots, x_{r-1})$ (see formula 4.3) which unambiguously represent the particular SGA populations if the parameter μ is fixed. The frequency vector entity x_i expresses the portion of the population x occupied by individuals of the binary code which is equal to $i \in \mathbb{Z}$. The set of states X_μ is contained in the unit $(r-1)$-dimensional simplex $\Lambda^{r-1} \subset \mathbb{R}^r$ (see formula 4.4).

We will also study the idealized, limit case of SGA in which the genetic universum Ω is finite ($r < +\infty$) but populations are infinite ($\mu = \#\mathbb{Z}$), for which the space of states is also infinite $\mathcal{E} = \Lambda^{r-1}$ (see Remark 4.3).

If the mutation and crossover parameters p_m, p_c and crossover type vector *type* (see Section 3.5) are constant then the Simple Genetic Algorithm that processes finite populations may be modeled by the uniform Markov chain with the space of states X_μ. The Markov kernel (transition function τ) may be characterized by the $n \times n$ transition probability matrix $\mathbf{Q}$.

Let us denote by $\pi^0_\mu \in \mathcal{M}(X_\mu)$ the n-dimensional probabilistic vector that characterize the sampling of the initial population P_0 which is associated with the frequency vector $x^0 \in X_\mu$. The Markov transition rule implies that

$$\pi^{t+1}_\mu = \mathbf{Q}\pi^t_\mu, \; t = 0, 1, 2, \ldots \tag{4.12}$$

where $\pi^t_\mu \in \mathcal{M}(X_\mu), t = 1, 2, \ldots$ is the n-dimensional probabilistic vector whose entries $(\pi^t_\mu)_x$ define the probability of the state $x \in X_\mu$ (as well as population unambiguously assigned to the state x) occurrence in the epoch t.

Genetic operator

In the case of the binary genetic universum Ω of the final cardinality $r < +\infty$ the fitness function $f : \Omega \to [0, M], M < +\infty$ is represented by the vector of its values

$$f \sim (f_0, f_1, \ldots, f_{r-1}) \in \mathbb{R}^r; \; f_i = f(i), i \in \Omega. \tag{4.13}$$

We will denote by diag(f) the $r \times r$ diagonal matrix, whose diagonal consists of f entries.

Definition 4.11. *The proportional selection operator is the mapping*

$$F : \Lambda^{r-1} \to \Lambda^{r-1};\; F(x) = \frac{\operatorname{diag}(f)\, x}{(f, x)}.$$

□

The scalar product (f, x) appearing in the above formula represents the mean fitness of the population represented by the state $x \in \Lambda^{r-1}$.

Remark 4.12. Let $P = (\Omega, \eta)$ be the arbitrary SGA population represented by the state $x \in X_\mu$, then

$$\forall i \in \Omega \;\; sel_f^3(\{i\}) = (F(x))_i$$

where sel_f^3 is the proportional selection distribution defined by formulas 3.29, 3.30. □

The above observation shows us that the probability of selecting the individual with the genotype $i \in \Omega$ from the population represented by the state $x \in X_\mu$ is equal to the ith coordinate of the value of the proportional selection operator $F(x)$. It is enough to divide both numerator and denominator at the right hand side of the formula 3.30 by the population cardinality μ in order to justify this observation. It also justifies the following simple remark.

Remark 4.13. Each value of the proportional selection operator $F(x), x \in \Lambda^{r-1}$ is the probability distribution on the genetic universum Ω ($\forall i \in \Omega \; 0 \leq (F(x))_i \leq 1,\; \sum_{i \in \Omega} (F(x))_i = 1 \;\forall x \in \Lambda^{r-1}$), so this operator may also be handled as the mapping

$$F : \Lambda^{r-1} \to \mathcal{M}(\Omega).$$

□

Next, we define the *mixing operator* that comprehends the results of mutation and crossover genetic operations.

Definition 4.14. *The symmetric matrix* $\mathbf{M} = \{m_{i,j}(\{0\})\},\; i, j \in \Omega$ *of the dimension* $r \times r$ *whose entries are defined by formulas 3.48, 3.49 will be called the mixing matrix.* □

Definition 4.15. *The mixing operator is the mapping* $M : \Lambda^{r-1} \to \Lambda^{r-1}$ *given by the formula*

$$(M(x))_i = (\sigma_i\, x)^{\mathrm{T}} \mathbf{M} \sigma_i\, x, \;\; x \in \Lambda^{r-1},\; i \in \Omega$$

where σ_i *stands for the* $r \times r$ *dimension permutation matrix with the entries* $(\sigma_i)_{j,k} = [j \oplus k = i],\; i, j, k \in \Omega$. □

Definition 4.16. *The genetic operator is the mapping $G : \Lambda^{r-1} \to \Lambda^{r-1}$, which is a composition of both: proportional selection and mixing operators*

$$G = M \circ F.$$

□

Remark 4.17. Proportional selection, mixing and genetic operators are continuously differentiable on the unit simplex Λ^{r-1} i.e. $F, M, G \in \mathrm{C}^1(\Lambda^{r-1})$. This means that they belong to the class $\mathrm{C}^1(A)$, where A is an open set in $\mathbb{R}^r$ so that $\Lambda^{r-1} \subset A$ and $0 \notin A$. □

Let us study the extensions of proportional selection mixing and genetic operators to the whole $\mathbb{R}^r$ in order to justify the above remark.

Remark 4.18. The extension of the mixing operator M is continuously differentiable in the whole $\mathbb{R}^r$ because it is simply the quadratic form of its argument. The extension of the proportional selection operator F is continuously differentiable in $\mathbb{R}^r$ except for zero, so this same range of differentiability is preserved for their composition $G = M \circ F$. □

Because the distance between Λ^{r-1} and 0 is strictly positive, we may easily select the open set A that contains Λ^{r-1} and does not contain 0 which is necessary to justify the thesis of Remark 4.17.

Theorem 4.19. *Let $x^t \in \Lambda^{r-1}$ be the frequency vector of the SGA population in the genetic epoch $t \geq 0$, then the i-th coordinate $(G(x^t))_i$ of the genetic operator value stands for the sampling probability of the individual with the genotype $i \in \Omega$ for the next epoch population P_{t+1}.* □

The above thesis may be drawn from the structuring process of the genetic operator whose coordinates are based on probability selection and mixing distributions 3.29, 3.30, 3.48, 3.49 and the symmetry condition delivered by Theorem 3.13. A detailed justification of the thesis of Theorem 4.19 can be found in the Vose monograph [193].

Let us consider two SGA populations in the consecutive genetic epochs P_t, P_{t+1} and corresponding frequency vectors $x^t, x^{t+1} \in \Lambda^{r-1}$. The state vector x^{t+1} produced by the algorithm may be interpreted as the random variable with the probability distribution π^{t+1} which depends only on x^t with respect to the Markov condition 4.6 for the family $\{P_t\}, t = 0, 1, 2, \ldots$.

Theorem 4.20. *(see Vose [193], Theorem 3.3) Let $x^t \in \Lambda^{r-1}$ be the frequency vector of the SGA population in the genetic epoch $t \geq 0$, then the expected frequency vector in the next epoch $t+1$ equals $G(x^t)$ (we can write using a short form $E(x^{t+1}) = G(x^t)$).* □

The next theorem determines the important relation between the genetic operator value and the Markov transition function (represented by the probability transition matrix) associated with the Simple Genetic Algorithm.

Theorem 4.21. *(see Nix and Vose [116]) If the parameters of SGA genetic operations: mutation rate, crossover rate and type $p_m, p_c, type$ (see Sections 3.5.1, 3.5.2) are constant (do not depend on the epoch number t), then the $n \times n$ transition probability matrix* $\mathbf{Q}$ *may be computed by the formula*

$$\mathbf{Q} = \{(\mathbf{Q})_{x,y}\}_{x,y \in X_\mu}, \ (\mathbf{Q})_{x,y} = \mu! \prod_{j=0}^{r-1} \frac{((G(x))_j)^{\mu y_j}}{(\mu y_j)!}.$$

□

The above theorem allows the effective computation of the transition probability matrix entries $\{(\mathbf{Q})_{x,y}\}, x, y \in X_\mu$ if the fitness vector $f = (f_0, \ldots, f_{r-1})$, population cardinality μ and binary code length l as well as genetic operator parameters $p_m, p_c, type$ are given. Additional information and formulas necessary for effective evaluation of the genetic operator G are contained in definitions 4.11–4.16.

The genetic operator plays a crucial role in the analysis of the Simple Genetic Algorithm and the above theorems are the first examples of its application. The next results will be presented in the following sections.

Remark 4.22. As far as the genetic operator G is completely determined by the parameters of the particular SGA instance (binary code length l, fitness vector $f = (f_0, \ldots, f_{r-1})$, mutation rate p_m, crossover rate and type $p_c, type$) it may correspond to an infinite, but countable class of SGA instances, which differs only in the population size $\mu \in \mathbb{N}$ or in the starting population P_0. Such a class of SGA instances will be called *associated with* or *spanned by the genetic operator G*. □

The Simple Genetic Algorithm as a dynamic semi-system

Definition 4.23. *(see Pelczar [125]) The dynamic semi-system is a triple (T, B, ϕ), where T is the topological space, $(B, +)$ is the topological alternating semi-group (there is no opposite element) and $\phi : T \times B \to T$ the mapping that satisfies the following conditions:*

1. *$\phi(\cdot, e) = I(\cdot)$ where $e \in B$ is the neutral element in B and I stands for the identity mapping on T,*
2. *$\forall p, t \in B, \forall x \in T \ \phi(\phi(x, p), t) = \phi(x, p + t)$,*
3. *the mapping ϕ is continuous with respect to both variables.* □

Let us first discuss the SGA with finite populations $\{P_t\}, t = 0, 1, 2, \ldots$ of the cardinality $\mu < +\infty$. The frequency vector of P_t will be denoted by $x^t \in X_\mu \subset \Lambda^{r-1}$ for $t = 0, 1, 2, \ldots$. The Markov transition function turns back the probabilistic measure $\tau(x^t) = \pi_\mu^{t+1} \in \mathcal{M}(X_\mu)$ like in the formula contained in Remark 4.7. Because X_μ is finite then also π_μ^t are vectors of the finite dimension $n = \#X_\mu$ (see Remark 4.2). The vector π^t_μ may also be handled as a discrete measure which is an element of the space $\mathcal{M}(\Lambda^{r-1})$, concentrated on the discrete set of points $X_\mu \subset \Lambda^{r-1}$.

Using the formula recursively $\pi^{t+1}_\mu = \mathbf{Q}\pi^t_\mu$ (see 4.12) we obtain

$$\forall t, p \in \mathbb{Z}_+ \;\; \pi^{t+p}_\mu = \mathbf{Q}^p \pi^t_\mu = \mathbf{Q}^{t+p}\pi^0_\mu. \tag{4.14}$$

Moreover, we have

$$\mathbf{Q}^{t+p} = \mathbf{Q}^t\,\mathbf{Q}^p = \mathbf{Q}^p\,\mathbf{Q}^t \;\; \forall t, p \in \mathbb{Z}_+. \tag{4.15}$$

Let us now set $T := \mathcal{M}(X_\mu)$ which is a set of n-dimensional probabilistic vectors with the topology induced from $\mathbb{R}^n$. We may also equivalently set T as the space of discrete probabilistic measures on Λ^{r-1} concentrated in the points from X_μ. Let B be the semi-group of transformations $T \to T$ spanned by the consecutive iterates of the matrix $\mathbf{Q}$, i.e. $\{\mathbf{Q}^p\}, p = 0, 1, 2, \ldots$ with the mapping composition as the group operation " + ". Each iterate $\mathbf{Q}^p$ is the continuous mapping of T to itself as the linear mapping of the finite dimensional vector-topological space. Now we can define

$$\phi : T \times B \ni (x, \mathbf{Q}^p) \to \mathbf{Q}^p\, x \in T$$

where $\mathbf{Q}^p, p \in \mathbb{Z}_+$ represents the arbitrary element of the semi-group B. The continuity condition for ϕ with respect to the first variable was already proved. Continuity with respect to the second variable is trivial because B is a discrete set. Moreover, ϕ satisfies condition 2 of the definition 4.23 because

$$\phi(\phi(x, \mathbf{Q}^p), \mathbf{Q}^t) = \mathbf{Q}^t\,(\mathbf{Q}^p\, x) = \mathbf{Q}^{t+p}\, x = \phi(x, \mathbf{Q}^{t+p}). \tag{4.16}$$

We have justified then the following remark:

Remark 4.24. The Simple Genetic Algorithm that transforms finite populations of the cardinality $\mu < +\infty$ may be modeled as a dynamic semi-system whose states belong to the space of discrete probabilistic measures on Λ^{r-1} concentrated in points from X_μ. □

If the Simple Genetic Algorithm processes finite populations ($\mu < +\infty$), then the transition from the current state $x \in X_\mu$ to the state $y \in X_\mu$ in the next genetic epoch is performed with the probability $(\mathbf{Q})_{x,y}$. If μ tends to infinity then X_μ becomes dense in Λ^{r-1} (see Remark 4.3) and each frequency vector $x \in \Lambda^{r-1}$ may represent the infinite population. The next epoch population represented by $x^{t+1} \in \Lambda^{r-1}$ that follows $x^t \in \Lambda^{r-1}$ in the infinite population SGA is obtained by infinite sampling with the probability distribution depending only on x^t.

What does the Markov transition rule $\tau(x^t)$ look like in this case? The answer may be found in the following theorem.

Theorem 4.25. *(see Vose [193], Theorem 13.2)* $\forall K > 0, \forall \varepsilon > 0, \forall \nu < 1$ $\exists N > 0$ *independent of* $x^0_\mu \in \Lambda^{r-1}$ *so that* $\forall\, 0 \le t \le K$

$$\Pr\left\{\mu > N \Rightarrow \left\|x^t_\mu - G^t(x^0_\mu)\right\| < \varepsilon\right\} > \nu$$

where $x^t_\mu \in \Lambda^{r-1}$ *is the frequency vector of the population* P_t *of the cardinality* μ *produced by the SGA and* $x^0_\mu \in \Lambda^{r-1}$ *stands for the frequency vector of the initial population* P_0 *of this same cardinality.* □

The above thesis may be interpreted as follows. When the number of the SGA population is sufficiently large then the population x^t will be followed by $x^{t+1} = G(x^t)$ with a probability which is arbitrarily close to one. The transition rule becomes deterministic, i.e. it will be simply the mapping $\Lambda^{r-1} \to \Lambda^{r-1}$.

Let us now set $T := \Lambda^{r-1} \subset \mathbb{R}^r$ with the induced topology and $B := \{G^p\}, p = 0, 1, 2, \ldots$ where $G^0 = I$ is the identity mapping on Λ^{r-1}. Each element of B of the form $G^p, p \in \mathbb{Z}_+$ is a continuous mapping on T (see Remark 4.17), then B constitutes the semi-group of continuous mappings with the mapping composition as the group operation "+". Because $G^p\, G^k = G^k\, G^p = G^{(p+k)}$ for arbitrary integers $p, k \ge 0$ then B is the alternating semi-group of mappings. Setting $\phi(x, G^p) := G^p(x), x \in T, G^p \in B$ we obtain the mapping which is continuous in both variables. The continuity with respect to the first variable is the simple matter of G continuity (see once more Remark 4.17). The continuity with respect to the second variable is trivial in the discrete topology on B. Moreover, we have:

$$\phi(\phi(x, G^p), G^k) = G^k(G^p(x)) = G^{k+p}(x) = \phi(x, G^{k+p}) \tag{4.17}$$

which completes the proof of both conditions appearing in definition 4.23 and the following remark may be drawn:

Remark 4.26. (see also Grygiel [80]) The Simple Genetic Algorithm with the infinite population ($\mu = +\infty$) may be modeled as the dynamic semi-system whose states belong to Λ^{r-1}. □

The above discussion as well as Theorem 4.25 also leads to another observation:

Remark 4.27. The initial range of the trajectory $\{x^t\}, t = 0, \ldots, K, K < +\infty$ of the finite population ($\mu < +\infty$) Simple Genetic Algorithm is located arbitrarily close to the trajectory of the infinite population SGA with an arbitrarily large probability if the population cardinality μ is sufficiently large. □

The considerations presented in this section show what kind of objects can be transformed regularly by the Simple Genetic Algorithm. They are probabilistic measures from the space $\mathcal{M}(X_\mu)$ in the case of finite populations of the size μ and in infinite populations their representations from the set

Λ^{r-1}. The transformation rule is static in both cases (does not depend on the genetic epoch counter t) and deterministic if $\mu = +\infty$. It is represented by the constant transition probability matrix $\mathbf{Q}$ in the first case and the genetic operator G in the second one.

Asymptotic results

Lemma 4.28. *(see Vose [193], Theorem 4.7) If the mutation rate is strictly positive ($p_m > 0$) then the genetic operator is strictly positively defined i.e. $(G(x))_i > 0, \forall i \in \Omega, \forall x \in \Lambda^{r-1}$.* □

It is enough to observe that if $p_m > 0$ then the mixing matrix $\mathbf{M}$ has strictly positive entries $m_{i,j}(\{0\}) > 0, i,j \in \Omega$ in order to justify the above theorem. It also means that, starting from the arbitrary population $x \in \Lambda^{r-1}$, another arbitrary population $y \in \Lambda^{r-1}$ may be reached in a single iteration step of the Simple Genetic Algorithm with a positive probability in this case. In other words, such a feature is caused by mutation which ensures the passage between two arbitrary states $x, y \in \Lambda^{r-1}$ with a positive probability when $p_m > 0$. Lemma 4.28 leads to another important observation:

Remark 4.29. The probability of producing an individual with an arbitrary genotype $i \in \Omega$ is strictly positive if the mutation rate is strictly positive ($p_m > 0$). This feature does not depend on the number of the genetic epoch. □

Moreover, combining Lemma 4.28 with the formula contained in Theorem 4.21 we can obtain:

Remark 4.30. If the mutation rate is strictly positive ($p_m > 0$) then the transition probability matrix $\mathbf{Q}$ is also strictly positive i.e. $(\mathbf{Q})_{x,y} > 0, \forall x, y \in X_\mu$. □

Now we are able to discuss the first result concerning SGA asymptotic behavior.

Theorem 4.31. *If the mutation rate is strictly positive ($p_m > 0$) then the Markov chain describing the finite population SGA ($\mu < +\infty$) is ergodic and a weak limit $\pi_\mu \in \mathcal{M}(X_\mu)$ exists, so that*

$$\lim_{t \to +\infty} \pi_\mu^t = \lim_{t \to +\infty} \mathbf{Q}^t \, \pi_\mu^0 = \pi_\mu$$

for the arbitrary initial measure $\pi_\mu^0 \in \mathcal{M}(X_\mu)$. □

The first part of the above thesis (the ergodicity) is the immediate conclusion of Remark 4.30 and the second may be drawn from the ergodic Theorem (see Feller [65]).

One important outcome of the Markov chain ergodicity is that the SGA will visit all states $x \in \Lambda^{r-1}$, then as a result will search all the set of phenotypes $\mathcal{D}_r$ regardless of the initial population P_0. Therefore ergodicity guarantees that the SGA becomes the well-defined global optimization algorithm if $p_m > 0$.

In particular we have:

Remark 4.32. If the mutation rate is strictly positive ($p_m > 0$) then the Simple Genetic Algorithm is asymptotically correct in the probabilistic sense (see definition 2.16). Moreover, it has the asymptotic guarantee of success (see definition 2.17). □

In other words, each local and global extreme will appear in at least one population produced by the SGA if the mutation rate is strictly positive and the number of genetic epochs is sufficiently large.

The next two theorems deliver information about the behavior of SGA state probability distributions when the number of individuals in the population grows to infinity.

Theorem 4.33. *(see Vose [193], Nix and Vose [116]) If the population cardinality grows to infinity ($\mu \to +\infty$) then the sequence of limit measures $\{\pi_\mu\} \subset \mathcal{M}(X_\mu)$ defined by the thesis of Theorem 4.31 contains the subsequence $\{\pi_{\mu_k}\} \subset \{\pi_\mu\}$ that weakly converges to some measure $\pi^* \in \mathcal{M}(\Lambda^{r-1})$.* □

The proof of this important feature is based on the Prochorow Theorem (see Feller [65]).

Definition 4.34. *The genetic operator $G : \Lambda^{r-1} \to \Lambda^{r-1}$ will be called focusing if for all $x \in \Lambda^{r-1}$ the sequence*

$$x, G(x), G^2(x), G^3(x), \ldots$$

converges in Λ^{r-1}. □

Let $w \in \Lambda^{r-1}$ be the limit of G iterates for some starting point $x \in \Lambda^{r-1}$. The continuity of G guarantees that:

$$G(w) = G\left(\lim_{t \to +\infty} G^t(x)\right) = \lim_{t \to +\infty} G^{t+1}(x) = w \tag{4.18}$$

then w is also the fixed point of G. Let us denote by

$$\mathcal{K} = \{w \in \Lambda^{r-1};\ G(w) = w\} \tag{4.19}$$

the set of all fixed points of the genetic operator G. If G is focusing, then $\mathcal{K} \neq \emptyset$ and $\mathcal{K}$ constitute the attractor of the dynamic semi-system associated with the infinite population SGA.

The final important theorem presented in this section determines the value of the limit measure π^* on the fixed set $\mathcal{K}$.

Theorem 4.35. *(see Nix and Vose [116], Theorem 3) If the genetic operator $G : \Lambda^{r-1} \to \Lambda^{r-1}$ associated with the class of Simple Genetic Algorithms is focusing and $\mathcal{K} \subset \Lambda^{r-1}$ is the set of its fixed points, then $\pi^*(\mathcal{K}) = 1$.* □

The above thesis may be commented upon as follows: if the genetic operator is focusing, then the infinite population SGA will oscillate among the fixed points of its genetic operator after a sufficiently large number of genetic epochs. Other states are achieved with a probability equal to zero.

Fixed points of the genetic operator and their stability

Let us start with the most convenient definition of fixed point stability of the iterating system on the SGA space of states Λ^{r-1}.

Definition 4.36.

1. *The fixed point $w \in \Lambda^{r-1}$ of the mapping $g : \Lambda^{r-1} \to \Lambda^{r-1}$ is stable (in the Lapunov sense) if, and only if, for each neighborhood $U_1; w \in U_1$ another neighborhood $U_2; w \in U_2$ exists so that for each starting point $y \in U_2$ the trajectory $\{y, g(y), g^2(y), \ldots\}$ lies in U_1 ($\{y, g(y), g^2(y), \ldots\} \subset U_1$).*
2. *The fixed point $w \in \Lambda^{r-1}$ of the mapping $g : \Lambda^{r-1} \to \Lambda^{r-1}$ is asymptotically stable if there is the neighborhood $U; w \in U$ so that for each $y \in U$ the trajectory $\{y, g(y), g^2(y), \ldots\}$ converges to w.* □

The following characterization may be drawn for fixed points of focusing, continuously differentiable mappings.

Theorem 4.37. *Let $w \in \Lambda^{r-1}$ be the fixed point of the continuously differentiable mapping $g \in \mathrm{C}^1(\Lambda^{r-1} \to \Lambda^{r-1})$. If the spectral radius (the maximum eigenvalue) of the differential $Dg|_w$ is greater than 1, then w is unstable. If the spectral radius of the differential $Dg|_w$ is less than 1, then w is asymptotically stable. The fixed point w is called hyperbolic if neither eigenvalue of the differential $Dg|_w$ equals 1.* □

Let us recall that the genetic operator $G : \Lambda^{r-1} \to \Lambda^{r-1}$ (see definition 4.16) is the composition of the proportional selection operator $F : \Lambda^{r-1} \to \Lambda^{r-1}$ (see definition 4.11) and the mixing operator $M : \Lambda^{r-1} \to \Lambda^{r-1}$ (see definition 4.15). Moreover, M can be extended to the continuously differentiable operator $\mathbb{R}^r \to \mathbb{R}^r$ while G and F can be extended to continuously differentiable operators from $\mathbb{R}^r \setminus \{0\}$ to $\mathbb{R}^r$ (see Remark 4.18).

In order to study the stability of the genetic operator it is reasonable to study the stability of its components.

Lemma 4.38. *(see Vose [193], Theorem 7.1)*
The differential $DF|_y \in \mathcal{L}(\mathbb{R}^r \to \mathbb{R}^r)$ of the extension of the proportional selection operator $F : \mathbb{R}^r \setminus \{0\} \to \mathbb{R}^r$ computed at $y \in \mathbb{R}^r \setminus \{0\}$ is given by the formula

$$DF|_y = \left(I - \frac{1}{(f,x)} f \cdot x\right) \frac{\operatorname{diag}(f)}{(f,x)}\Bigg|_{x=y}$$

where $f \cdot x$ denotes the matrix $\{f_i x_j\}$ of the dimension $r \times r$ which is the tensor product of the vectors f and x. □

Let us assume, for a while, that the fitness function $f : \Omega \to [0, M]$ is injective i.e. $f_i \neq f_j,\ i \neq j,\ i,j \in \Omega$. In this case we have

$$\exists!\ \theta \in \Omega;\ f_\theta = f(\theta) = \max\{f_j, j \in \Omega\}. \tag{4.20}$$

If we assume, moreover, that fitness is non trivial i.e. $f_\theta > 0$ then two simple observations are valid (see Vose [193]):

Remark 4.39. (see Vose [193], Theorem 10.3) The spectrum (set of eigenvalues) of the differential operator $DF|_{x_\theta}$, $x_\theta = (0, \ldots, 0, \overset{(\theta)}{1}, 0, \ldots, 0)$ equals

$$\operatorname{spec}(DF|_{x_\theta}) = \left\{\frac{f_i}{f_\theta};\ 0 \leq i \leq r-1, i \neq \theta\right\}$$

□

Remark 4.40. (see Vose [193], Theorem 10.4) If the fitness function $f : \Omega \to [0, M]$ is injective and non trivial then x_θ is the only stable fixed point of the proportional selection operator F. Other fixed points of F are hyperbolic. □

The next part of this section will be devoted to the fixed points of the mixing operator $M : \Lambda^{r-1} \to \Lambda^{r-1}$.

Theorem 4.41. *(see Vose [193], Theorem 6.13, also Kołodziej [94]) The differential of the extension of the mixing operator $DM|_y \in \mathcal{L}(\mathbb{R}^r \to \mathbb{R}^r)$ computed at $y \in \mathbb{R}^r$ is given by the formula*

$$DM|_y = 2 \sum_{k=0}^{r-1} \sigma_k^{-1}\, \mathbf{M}_*\, \sigma_k\, x_k \Bigg|_{x=y}$$

where σ_k is the $r \times r$ dimension permutation matrix with entries $(\sigma_i)_{j,k} = [j \oplus k = i],\ i,j,k \in \Omega$, and $\mathbf{M}_*$ is the $r \times r$ dimension matrix of entries $(\mathbf{M}_*)_{i,j} = (\mathbf{M})_{i \oplus j, j},\ i,j \in \Omega$ and $(\mathbf{M})_{i,j},\ i,j \in \Omega$ are entries of the mixing matrix $\mathbf{M}$ (see formula 4.14). □

Remark 4.42. The derivative of the extension of the mixing operator

$$\mathbb{R}_r \ni x \to \left\{2 \sum_{k=0}^{r-1} \sigma_k^{-1}\, \mathbf{M}_*\, \sigma_k\, x_k\right\} \in \mathcal{L}(\mathbb{R}^r \to \mathbb{R}^r)$$

is a linear mapping. □

Let us study the features of the matrix operator $\mathbf{M}_*$ as the introductory step to studying fixed points of the operator M.

Theorem 4.43. *(see Vose [193], Theorem 6.3)*

1. *The maximum eigenvalue of the matrix $\mathbf{M}_*$ equals 1 and corresponds to the left eigenvector $\mathbf{1}$ which is perpendicular to the unit simplex Λ^{r-1} in $\mathbb{R}^r$.*
2. *If the mutation rate is strictly positive ($p_m > 0$) then other eigenvalues of the matrix $\mathbf{M}_*$ have their modules less than 0.5.* □

The above theorem partially motivates the next theses.

Theorem 4.44. *(see Vose [193], Theorem 6.13)*

1. $\operatorname{spec}(DM|_x) = 2(\mathbf{1}, x)\operatorname{spec}(\mathbf{M}_*) \;\; \forall x \in \mathbb{R}^r$
2. *The maximum eigenvalue of the differential $DM|_x$ equals $2(\mathbf{1}, x)$ and corresponds to the eigenvector $\mathbf{1}$ for all $x \in \mathbb{R}^r$. If we restrict the mapping M to the unit simplex Λ^{r-1}, then this eigenvalue has to be ignored in the stability analysis, because $\mathbf{1}$ is perpendicular to Λ^{r-1} and does not affect the behavior of the operator M on Λ^{r-1}.*
3. *If the mixing operator M is strictly understood, according to the definition 4.15, as the mapping $\Lambda^{r-1} \to \Lambda^{r-1}$ then $\operatorname{spec}(DM|_x)$ does not depend on $x \in \Lambda^{r-1}$.* □

Studying the stability of the fixed points of the mixing operator M is much more complicated than in the case of the proportional selection operator F. We restrict ourselves only to features derived by Vose (see Vose [193], Theorem 10.8), which are important for future considerations.

Theorem 4.45. *If the mutation rate is strictly positive ($p_m > 0$), then the centroid of the simplex $\frac{1}{r}\mathbf{1} \in \Lambda^{r-1}$ is the fixed point of the mixing operator M.* □

Using the well known chain rule for differentiating the composition of functions we obtain:

$$DG|_x = DM|_{F(x)} \circ DF|_x \quad \forall x \in \mathbb{R}^r \setminus \{0\}. \tag{4.21}$$

A more detailed formula for DG is also delivered by Proposition 2.3 in [195].

The vertices $e_0, \dots, e_{r-1}$, $e_i = (0, \dots, 0, \overset{(i)}{1}, 0, \dots, 0)$, $i = 0, \dots, r-1$ of the simplex Λ^{r-1} are analyzed as potential fixed points of the genetic operator G by Vose and Wright [195]. Their results may be stressed as follows:

Remark 4.46. If the mutation vanishes ($p_m = 0$) then the vertices $e_0, \dots, e_{r-1}$ of the simplex Λ^{r-1} are fixed points of the genetic operator G. □

Each vertex $e_i, i \in \Omega$ represents the monochromatic population P that contain only individuals of the single genotype $i \in \Omega$ (see Remark 4.4). Neither selection nor crossover can modify the genotype of the individual from P, so the expected population that follows e_i is also $G(e_i) = e_i$.

The stability of the Λ^{r-1} vertices may be analyzed in the above case by analyzing the spectral radius of $DG|_{e_k}$.

Theorem 4.47. *(see Vose and Wright [195], Theorem 3.4) If there is no mutation ($p_m = 0$) then*

$$\mathrm{spec}(DG|_{e_k}) = \left\{ \frac{f_{i\oplus k}}{f_k} \sum_{k=0}^{r-1} (\eta_k + \eta_{\hat{k}})[k \oplus i = 0], i = 1, \ldots, r-1 \right\} \cup \{0\}$$

□

Moreover, Vose and Wright suggested in [195] that Λ^{r-1} vertices are the only candidates for stable fixed points of G if the mutation vanishes in SGA ($p_m = 0$) (see [195], Conjecture 4.4).

Next, we will study the behavior of the genetic operator G in cases where there is no crossover ($p_c = 0$) and mutation is positive ($p_m > 0$). This case has been studied by Grygiel [80]. The mixing component M of the genetic operator is reduced to the linear operator $\mathbb{R}^r \to \mathbb{R}^r$ with the matrix $\{\tilde{m}_{i,j}\}$ of coefficients

$$\tilde{m}_{i,j} = p_m^{(\mathbf{1}, i\oplus j)} (1 - p_m)^{l - (\mathbf{1}, i\oplus j)}. \tag{4.22}$$

The genetic operator can be expressed by the formula

$$G(x) = M(F(x)) = \frac{1}{(f,x)} \mathbf{H}\, x, \; x \in \Lambda^{r-1} \tag{4.23}$$

where $\mathbf{H} = \{\tilde{m}_{i,j} f_j\}$ is the $r \times r$ matrix.

Theorem 4.48. *(see Grygiel [80], Proposition 2) If the SGA genetic operations are restricted only to mutation ($p_c = 0$, $p_m > 0$) then there is exactly one fixed point w of the genetic operator G. The fixed point w has strictly positive coordinates ($w \in \mathrm{int}(\Lambda^{r-1})$). Moreover, w is the eigenvector which corresponds to the maximum eigenvalue λ_{max} of the matrix $\mathbf{H}$ i.e. ($\mathbf{H}\, w = \lambda_{max}\, w$). The fixed point w is asymptotically stable, moreover, the whole simplex Λ^{r-1} is the attractor of w, i.e.*

$$\forall x \in \Lambda^{r-1} \; \lim_{t \to +\infty} G^t(x) = w.$$

□

The last problem discussed in this section will be the characteristic of hyperbolic fixed points of the genetic operator G.

Definition 4.49. *The genetic operator $G : \Lambda^{r-1} \to \Lambda^{r-1}$ will be called* regular *if for each set $C \subset \Lambda^{r-1}$ with a zero measure in Λ^{r-1} the set $G^{-1}(C)$ also has a zero measure in Λ^{r-1}. We will understand the measure in Λ^{r-1} as the measure on the $(r-1)$-dimensional hyperplane that contains Λ^{r-1}.* □

The two following theorems more precisely define the condition by which the genetic operator G is regular and invertible as well as inform us about the number of fixed points of such operators.

Theorem 4.50. *(see Vose [193], Theorems 9.3 and 13.6) If the crossover rate is strictly less than one ($p_c < 1$) and the mutation rate satisfies $0 < p_m < \frac{1}{2}$ then the genetic operator G is invertible on Λ^{r-1} and $G^{-1} \in C^1(\mathrm{int}(\Lambda^{r-1}))$. Moreover, G is regular on Λ^{r-1}.* □

Theorem 4.51. *(see Vose [193], Theorem 12.1) If the genetic operator G has only hyperbolic fixed points, then it has a finite number of fixed points in Λ^{r-1}.* □

An attempt to the evaluation the rate of convergence – logarithmic convergence

Definition 4.52. *The genetic operator $G : \Lambda^{r-1} \to \Lambda^{r-1}$ will be called logarithmic convergent if*

$$\forall \rho \in \mathcal{M}(\Lambda^{r-1}),\ \forall \varepsilon > 0\ \exists A \subset \Lambda^{r-1};\ \rho(A) = 1 - \varepsilon$$

so that

$$\forall x \in A,\ \forall \delta \in (0,1)\ \exists k \geq 1; k = O(-\log(\delta))$$

(k converges with the same order as $(-\log(\delta))$), moreover,

$$\left\| G^k(x) - \omega(x) \right\| < \delta, \text{ where } \omega(x) = \lim_{t \to +\infty} G^t(x)$$

□

If the particular genetic operator G is logarithmic convergent, then we can evaluate the rate of convergence of the infinite population SGA which starts from the set A and tends to the fixed point $\omega(x), x \in A$. The set A may have the measure arbitrarily close to the measure of the whole simplex Λ^{r-1}.

The logarithmic convergence of the genetic operator G holds under the following conditions.

Theorem 4.53. *(see Vose [193], Theorem 13.10) If the genetic operator $G : \Lambda^{r-1} \to \Lambda^{r-1}$ is focusing and regular (see definitions 4.34, 4.49) and all the fixed points of G are hyperbolic, then G is logarithmic convergent.* □

The approximation of the fixed points of the genetic operator

Let us consider the class of instances of the Simple Genetic Algorithm spanned by the single genetic operator G (see Remark 4.22). In this short section we try to answer how close finite populations, generated by the above algorithms during a finite number of genetic epochs, can approximate fixed points of their spanning genetic operator.

We restrict ourselves to the case in which G is focusing and the set $\mathcal{K}$ of fixed points is finite, so all fixed points are isolated. Sufficient conditions for $\#\mathcal{K} < +\infty$ are delivered by the Theorem 4.51.

Theorem 4.54. *(see Telega [184], Cabib, Schaefer, Telega [43]) Let us assume that the genetic operator $G : \Lambda^{r-1} \to \Lambda^{r-1}$ is focusing and its set of fixed points is finite ($\#\mathcal{K} < +\infty$). Let us define*

$$\mathcal{K}_\varepsilon = \left\{x \in \Lambda^{r-1};\ \exists y \in \mathcal{K};\ d(x,y) < \varepsilon\right\}$$

being the open ε-envelope of $\mathcal{K}$ in the $(r-1)$-dimensional hyperplain that contains Λ^{r-1}, where $d(\cdot,\cdot)$ stands for the Euclidean distance in $\mathbb{R}^{r-1}$. We assume moreover, that mutation is strictly positive ($p_m > 0$), then

$$\forall \varepsilon > 0,\ \forall \eta > 0,\ \exists N \in \mathbb{N},\ \exists W(N) \in \mathbb{N};$$
$$\forall \mu > N,\ \forall k > W(N)\ \ \pi_\mu^k(\mathcal{K}_\varepsilon) > 1 - \eta.$$

where π_μ^k are measures associated with the infinite sub-class of Simple Genetic Algorithms spanned by G. □

In other words, if G is focusing, then sufficiently large SGA populations will concentrate close to the set $\mathcal{K}$ with an arbitrarily large probability $1-\eta$ after the sufficiently large number of genetic epochs. The above theorem will be intensively used when studying the asymptotic features of sampling measures presented in the next section.

The proof of Theorem 4.54 will be preceded by two lemmas and one technical remark concerning the relation between the support of measures π_μ, π_μ^k and the set $\mathcal{K}_\varepsilon$.

Lemma 4.55. *There is an infinite sub-class of Simple Genetic Algorithms spanned by G so that under the assumptions of Theorem 4.54*

$$\forall \varepsilon > 0,\ \forall \eta > 0\ \exists N \in \mathbb{N};\ \forall \mu > N\ \ \pi_\mu(\mathcal{K}_\varepsilon) > 1 - \eta$$

□

Proof. Let us consider the limit measures π_μ associated with the finite population SGA (see Theorem 4.31). According to Theorem 4.33 the sequence $\{\pi_\mu\}$ contains the infinite sub-sequence $\{\pi_{\mu_\xi}\}$ that converges to π^*. Let us select the sub-class of Simple Genetic Algorithms spanned by G whose measures $\{\pi_{\mu_\xi}^k\}$ converge to the elements π_{μ_ξ} of this sub-sequence if the number of genetic epochs tends to infinity ($k \to +\infty$). For the sake of simplicity we will denote by π_μ the elements of the sub-sequence π_{μ_ξ} in the remaining part of the proof.

Because the measure π^* is concentrated on $\mathcal{K}$ (see Theorem 4.35) then all the sets $\mathcal{K}_\varepsilon$ are π^*-continuous which means that $\pi^*(\partial\mathcal{K}_\varepsilon) = 0$ (see Billingsley [28], Section 29). The weak convergence $\pi_\mu \to \pi^*$ implies $\pi_\mu(\mathcal{K}_\varepsilon) \to \pi^*(\mathcal{K}_\varepsilon)$ (see Billingsley [28], Theorem 29.1), then for the arbitrary $\varepsilon > 0$ we have

$$\forall \eta > 0 \ \exists N; \ \forall \mu > N \ \ |\pi^*(\mathcal{K}_\varepsilon) - \pi_\mu(\mathcal{K}_\varepsilon)| < \eta$$

then also $1 - \pi_\mu(\mathcal{K}_\varepsilon) < 1$ because $\pi^*(\mathcal{K}_\varepsilon) = \pi^*(\mathcal{K}) = 1$ and $0 \leq \pi_\mu(\mathcal{K}_\varepsilon) \leq 1$ which completes the proof. □

Remark 4.56. Let us assume that $\mu \in \mathbb{N}$ is arbitrary and such, that the space of states $X_\mu \subset \Lambda^{r-1}$ for the finite population SGA exists then $\forall \varepsilon > 0 \ \exists \hat{\varepsilon} \in (0, \varepsilon]$ such, that:

1. $\mathcal{K}_{\hat{\varepsilon}}$ is π_μ-continuous and $\pi_\mu(\mathcal{K}_{\hat{\varepsilon}}) = \pi_\mu(\mathcal{K}_\varepsilon)$,
2. $\forall k \in \mathbb{N} \ \ \mathcal{K}_{\hat{\varepsilon}}$ is π_μ^k-continuous and $\pi_\mu^k(\mathcal{K}_{\hat{\varepsilon}}) = \pi_\mu^k(\mathcal{K}_\varepsilon)$. □

Proof. Let us recall that the measures π_μ and π_μ^k are concentrated on the finite set $X_\mu \subset \Lambda^{r-1}$ for all $k \in \mathbb{N}$. Let us define:

$$A = \mathcal{K}_\varepsilon \cap X_\mu, \ \ B = \bar{\mathcal{K}_\varepsilon} \cap X_\mu, \ \ C = B \setminus A = \partial\mathcal{K}_\varepsilon \cap X_\mu.$$

Observe that all the above sets A, B, C and $\mathcal{K}$ are finite. We may consider two separate cases:

1. $C = \emptyset$ which implies $\pi_\mu(\partial\mathcal{K}_\varepsilon) = 0$ and then the set $\mathcal{K}_\varepsilon$ is π_μ-continuous. We may set $\hat{\varepsilon} = \varepsilon$.
2. $\#C > 0$ then we may define

$$c = \max_{y \in \mathcal{K}_\varepsilon \cup X_\mu} \left\{ \min_{x \in \mathcal{K}} \{d(x, y)\} \right\}.$$

The constant c always exists, because both sets on which maximum and minimum are computed are finite. Obviously, from the definition of $\mathcal{K}_\varepsilon$ we get $c < \varepsilon$. Now it is enough to set $\hat{\varepsilon} \in (c, \varepsilon)$ because then $\overline{\mathcal{K}_{\hat{\varepsilon}}} \subset \text{int}(\mathcal{K}_\varepsilon)$ and $\partial\mathcal{K}_{\hat{\varepsilon}} \cap \partial\mathcal{K}_\varepsilon = \emptyset$ which implies $C \cap \partial\mathcal{K}_{\hat{\varepsilon}} = \emptyset$. Moreover, $\partial\mathcal{K}_{\hat{\varepsilon}} \cap A = \emptyset$ because $A \subset \text{int}(\mathcal{K}_{\hat{\varepsilon}})$. Finally, $\partial\mathcal{K}_{\hat{\varepsilon}} \cap X_\mu = \emptyset$ then $\pi_\mu(\partial\mathcal{K}_{\hat{\varepsilon}}) = 0$ which completes the proof of the first thesis of the remark.

The second thesis of the remark can be proved identically to the first one. □

Lemma 4.57. *If the assumptions of Theorem 4.54 hold, then*

$$\forall \varepsilon > 0 \ \forall \eta > 0 \ \forall \mu \in \mathbb{N} \ \exists K(\mu);$$

$$\forall k > K(\mu) \ \ \left|\pi_\mu(\mathcal{K}_\varepsilon) - \pi_\mu^k(\mathcal{K}_\varepsilon)\right| < \eta.$$

□

Proof. Similarly to the previous instance π_μ and π_μ^k may be handled as probabilistic measures defined on the $r-1$ hyperplane containing Λ^{r-1} that vanishes out of Λ^{r-1}. Let us select $\eta > 0$, $\mu \in \mathbb{N}$ and $0, \hat{\varepsilon} \leq \varepsilon$ so that $\mathcal{K}_{\hat{\varepsilon}}$ is π_μ-continuous and $\pi_\mu(\mathcal{K}_{\hat{\varepsilon}}) = \pi_\mu(\mathcal{K}_\varepsilon)$ (see remark 4.56). From Theorem 4.31 the sequence $\pi_\mu^k \to \pi_\mu$ weakly for $k \to +\infty$. Then there is $K(\mu)$ so that $\forall k > K(\mu)$ we have

$$\left|\pi_\mu(\mathcal{K}_{\hat{\varepsilon}}) - \pi_\mu^k(\mathcal{K}_{\hat{\varepsilon}})\right| < \eta.$$

Using remark 4.56 we directly obtain the thesis of Lemma 4.57. □

Proof of Theorem 4.54. Let us select $\varepsilon, \eta > 0$ and set $\varsigma = \frac{\eta}{2}$. From Lemma 4.55 we may derive

$$\exists N; \ \forall \mu > N \ \ \pi_\mu(\mathcal{K}_\varepsilon) > 1 - \varsigma = 1 - \frac{\eta}{2}.$$

For an arbitrary, proper $\mu \in \mathbb{N}$ we have from Lemma 4.57

$$\exists K(\mu); \ \forall k > K(\mu) \ \ \left|\pi_\mu(\mathcal{K}_\varepsilon) - \pi_\mu^k(\mathcal{K}_\varepsilon)\right| < \frac{\eta}{2}$$

which also implies that

$$\pi_\mu^k(\mathcal{K}_\varepsilon) - \pi_\mu(\mathcal{K}_\varepsilon) > -\frac{\eta}{2}$$

which added to the previous inequality yields

$$\pi_\mu^k(\mathcal{K}_\varepsilon) > 1 - \eta.$$

□

Asymptotic features of sampling measures

The considerations presented in some earlier sections (especially in *The Simple Genetic Algorithm as a dynamic semi-system* and *Asymptotic results*) show that the Simple Genetic Algorithm can regularly transform measures on its space of states Λ^{r-1}. Theorems 4.31, 4.33, 4.35 and 4.54 also show their asymptotic features.

Let us try to define when the Simple Genetic Algorithm can generate the sequence of sampling measures over the admissible set $\mathcal{D}$ which is convergent and its limit can be helpful in solving global optimization problem $\mathbf{\Pi_4}$ (see Section 2.1). We recall that problem $\mathbf{\Pi_4}$ consists of finding sets which are the central parts of the basins of attraction for the objective function Φ. Moreover, we try to formulate the condition that forces the SGA to satisfy the above need.

The requested sequence of sampling measures belonging to $\mathcal{M}(\mathcal{D})$ can be formulated in two steps:

Step I.

We will intensively use the following features of the genetic operator called SGA *heuristics* (see Vose [193]) originally defined as $G : \Lambda^{r-1} \to \Lambda^{r-1}$ (see definition 4.16, Theorems 4.19, 4.20, 4.25 and remark 4.26):

1. $G(x)$ is the expected population in the epoch that immediately follows the epoch in which the population vector $x \in \Lambda^{r-1}$ appears,
2. G is the evolutionary low of the abstract, infinite population SGA ($\mu = +\infty$),
3. Each coordinate $(G(x))_i, i = 0, 1, \dots, r-1$ stands for the sampling probability of the individual with the genotype $i \in \Omega$ in the epoch that immediately follows the epoch in which the population vector $x \in \Lambda^{r-1}$ appears.

Each point of the simplex $x \in \Lambda^{r-1}$ defines the unique measure $\theta(x) \in \mathcal{M}(\mathcal{D}_r)$ (see Section 3.1) so that the following one-to-one mapping may be formally established:

$$\begin{aligned} &\theta : \Lambda^{r-1} \to \mathcal{M}(\mathcal{D}_r); \\ &\forall x \in \Lambda^{r-1},\ \forall y \in \mathcal{D}_r,\ \forall i \in \Omega, \\ &y = \text{code}(i) \Rightarrow \theta(x)(\{y\}) = x_i \end{aligned} \tag{4.24}$$

The third feature of the genetic operator G allows us to define the mapping that turns back the sampling measure over the set of phenotypes:

$$\begin{aligned} &\hat{\theta} : \Lambda^{r-1} \to \mathcal{M}(\mathcal{D}_r); \\ &\forall x \in \Lambda^{r-1},\ \forall y \in \mathcal{D}_r,\ \forall i \in \Omega, \\ &y = \text{code}(i) \Rightarrow \hat{\theta}(\{y\}) = (G(x))_i \end{aligned} \tag{4.25}$$

Remark 4.58. Taking the mapping θ into account, the space $\mathcal{M}(\mathcal{D}_r)$ may be handled as the new space of states of the Simple Genetic Algorithm, in the same way as the simplex Λ^{r-1}. The role of heuristics will be played by the mapping:

$$\mathcal{M}(\mathcal{D}_r) \ni \sigma \to \hat{\theta}(\theta^{-1}(\sigma)) \in \mathcal{M}(\mathcal{D}_r).$$

□

The measures θ and $\hat{\theta}$ can also be identified with the other measures θ' and $\hat{\theta}'$ that belong to the space $\mathcal{M}(\mathcal{D})$. Both θ' and $\hat{\theta}'$ are defined over the whole admissible set $\mathcal{D}$ and concentrated at discrete phenotype points $\mathcal{D}_r \subset \mathcal{D}$. They will satisfy:

$$\theta'(x)(A) = \theta(x)(A \cap \mathcal{D}_r),\ \hat{\theta}'(x)(A) = \hat{\theta}(x)(A \cap \mathcal{D}_r)\ \ \forall x \in \Lambda^{r-1} \tag{4.26}$$

where $A \subset \mathcal{D}$ is an arbitrary set measurable in the Lesbegue sense.

Remark 4.59. Let us assume the finite population SGA ($\mu < +\infty$) and $x \in X_\mu$ is the vector of the population $P = (\Omega, \eta)$ in the particular genetic epoch, then $\theta'(x)$ is the counting measure, which means that for all $A \subset \mathcal{D}$ measurable in the Lesbegue sense

$$\theta'(x)(A) = \frac{1}{\mu} \sum_{y \in \mathrm{supp}(\eta)} \eta(y)\,[code(y) \in A].$$

□

Step II.

Now we introduce the mapping $\Psi : \mathcal{M}(\mathcal{D}_r) \to \mathcal{M}(\mathcal{D})$ that returns the special kind of measures which posses $L^p(\mathcal{D}), \mathcal{D} \subset \mathbb{R}^N$ density functions based on the probabilistic measure concentrated on the set of phenotypes $\mathcal{D}_r$ (see Schaefer, Jabłoński [154]).

Let $\vartheta_{(j)}$ be the domain of the inverse binary affine encoding defined by the formula 3.14 associated with the genotype $j = (j_1, \ldots, j_N) \in \Omega$. We recall that the substrings $\{j_i\}, j = 1, \ldots, N$ encode consecutive coordinates of the phenotype that belong to $\mathcal{D}_r \subset \mathbb{R}^N$. In the case of Gray encoding we assign $\vartheta_{(v(k))}$ to the genotype $k \in \Omega$, where $v : \Omega \to \Omega$ denotes the Gray encoding function (see 3.15).

If $\mathcal{D}$ has the Lipschitz boundary (see e.g. Zeidler [207]) then $\vartheta_{(j)}, j \in \Omega$ are measurable sets in the Lesbegue sense. Now we are prepared to set for the arbitrary measure $\omega \in \mathcal{M}(\mathcal{D}_r)$ and the Lesbegue measurable set $A \subset \mathcal{D}$

$$\Psi(\omega)(A) = \int_A \rho_\omega \, dx \tag{4.27}$$

where the right-hand-side integral is computed according to the Lesbegue measure. The density function is given by

$$\rho_\omega(x) = \sum_{j \in \Omega} \frac{\omega(\{\mathrm{code}_a(j)\})}{\mathrm{meas}(\vartheta_{(j)} \cap \mathcal{D})} \chi_{\vartheta_{(j)}}(x) \tag{4.28}$$

where $\chi_{\vartheta_{(j)}}$ denotes the characteristic function of the open brick $\vartheta_{(j)}$ that contains the phenotype $\mathrm{code}_a(j)$. The above construction follows the more general one presented in Section 3.1 (see formula 3.4). In the case of Grey encoding we can set

$$\rho_\omega(x) = \sum_{j \in \Omega} \frac{\omega(\{\mathrm{code}_G(k)\})}{\mathrm{meas}(\vartheta_{(v(k))} \cap \mathcal{D})} \chi_{\vartheta_{(v(k))}}(x). \tag{4.29}$$

Remark 4.60. If $\mathcal{D}$ has the Lipschitz boundary then $\rho_\omega \in L^p(\mathcal{D}),\ p \in [1, +\infty)$. □

Proof The density function ρ_ω is piecewise constant and bounded in $\mathcal{D}$. In particular it is constant in the sets $\{\vartheta_{(j)}\}, j \in \Omega$ which form the regular partitioning of $\mathcal{D}$ so that $\vartheta_{(i)} \cap \vartheta_{(j)} = \emptyset,\ i \neq j$ which was included in the partitioning construction (see Section 3.1). The partitioning forms the whole measure support $(\text{meas}\left(\bigcup_{i\in\Omega} \vartheta_{(i)}\right) = \text{meas}(\mathcal{D}))$ for this function. Such conditions guarantee the existence and boundedness of the Lesbegue integrals $\int_{\mathcal{D}} \rho_\omega^p$ for all $p \in \mathbb{N}$ which motivates the above thesis. □

Moreover, the following remark may be drawn from formulas 4.28, 4.29.

Remark 4.61. The mapping $\Psi : \mathcal{M}(\mathcal{D}_r) \to \mathcal{M}(\mathcal{D})$ is injective, in particular density ρ_ω is uniquely determined almost everywhere in $\mathcal{D}$ by the measure $\omega \in \mathcal{M}(\mathcal{D}_r)$. □

One simple but very important observation may also be performed for the introduced measures associated with the fixed point of the genetic operator.

Remark 4.62. If $x \in \mathcal{K}$ is the arbitrary fixed point of the genetic operator G, then $\theta(x) = \hat{\theta}(x),\ \theta'(x) = \hat{\theta}'(x)$ and $\Psi(\theta(x)) = \Psi(\hat{\theta}(x))$. □

Let us assume that the fixed points of the genetic operator G represent populations that contain maximum information about the global optimization problem to be solved, which can be gathered by any SGA algorithm spanned by G.

Now we are going to define the condition that may be necessary for successful solving of problem $\mathbf{\Pi_4}$ by using the SGA.

Let us denote by $\mathcal{W}$ the finite set of local maximizers to the objective function Φ, and by $\{\mathcal{B}_{x^+}\}, x^+ \in \mathcal{W}$ the family of their basins of attraction.

Definition 4.63. *We can say that the class of SGA spanned by the genetic operator G is well tuned to the set of local maximizers $\mathcal{W}$ if:*

1. *G is focusing and the set of its fixed points $\mathcal{K}$ is finite,*
2. *$\forall x^+ \in \mathcal{W}\ \exists C(x^+)$ closed set in $\mathcal{D}$ so that $x^+ \in C(x^+) \subset \mathcal{B}_{x^+}$, $\text{meas}(C(x^+)) > 0$ and*

$$\rho_{\theta(z)} \geq \textit{threshold} \quad x \in C(x^+)$$

$$\rho_{\theta(z)} < \textit{threshold} \quad x \in \mathcal{D} \setminus \bigcup_{x^+\in\mathcal{W}} C(x^+)$$

where $z \in \mathcal{K}$ is the arbitrary fixed point for G, $\theta(z)$ is the discrete measure associated with the fixed point z according to the formula 4.24. The positive constant threshold stands for the definition's parameter. □

In other words, the above definition postulates that the $L^P(\mathcal{D})$-regular measure densities $\rho_{\theta(z)}$ associated with all G fixed points z dominate almost everywhere on some central parts of the basins of attraction $C(x^+)$ of local maximizers x^+ which belong to the group of our special interest ($x^+ \in \mathcal{W}$), if the class of the SGA spanned by G is well tuned to $\mathcal{W}$.

Remark 4.64. The well tuning condition depends on the main objects of the optimization problem, such as the objective function Φ and the admissible domain $\mathcal{D}$, on the encoding quantities, such as the phenotype mesh $\mathcal{D}_r$ and the encoding function code : $\Omega \rightarrow \mathcal{D}_r$ and the SGA parameters $f, p_m, p_c, type$ (see Section 3.5). □

The above remark only provides an overall characterization of the SGA's well tuning. Really, the question of whether the Simple Genetic Algorithm is well tuned to the prescribed set of local maximizers of the arbitrary objective function is still open.

Adamska has studied experimentally the well tuning condition for some well-known benchmarks of global optimization. Her results were published in the paper [153]. Computational tests, as performed by Adamska, consist of finding the fixed point of the genetic operator and then comparing it with the objective function in order to find out if the particular SGA instance is well tuned, what the *threshold* parameter should be and what the relation of this parameter to the number of local maximizers to be involved in the well tuning condition is.

In order to effectively obtain the fixed point of the genetic operator, the SGA with only mutation is selected. In this case there exists the unique, stable fixed point w of the genetic operator which constitutes the eigenvector corresponding to the maximum eigenvalue of the $r \times r$ matrix $\mathbf{H} = \{\tilde{m}_{i,j} f_j\}$, where $\tilde{m}_{i,j} = p_m^{(\mathbf{1}, i \oplus j)} (1 - p_m)^{l - (\mathbf{1}, i \oplus j)}$ (see formulas 4.22, 4.23 and Theorem 4.48). The length of binary code was set at $l = 8$ while the mutation rate was $p_m = 0.05$ for all the cases presented below. The binary genetic universum Ω may be identified with the subset $\{0, 1, \ldots, 255\} \subset \mathbb{Z}_+$ of non-negative integers.

We will study the one-dimensional global optimization problems for which the objective function $\Phi : \mathcal{D} \rightarrow [0, M]$ is defined on the closed interval $\mathcal{D} = [left, right] \subset \mathbb{R}$. Fitness entries $\{f_i\}, i \in \Omega$ were computed immediately from the objective by using the affine encoding function $\text{code}_a : \Omega \rightarrow \mathcal{D}$ i.e.

$$f_i = \Phi(\text{code}_a(i)) = \Phi\left(left + i\,\frac{right - left}{2^l}\right), \quad i \in \Omega. \tag{4.30}$$

Eigenvalue/eigenvector linear algebra problems for the matrix $\mathbf{H}$ were solved by the symbolic processor MAPLE. The first example was associated with the one-dimensional Rastrigin benchmark function

$$\Phi_R(x) = -x^2 + \cos(2\pi\, x) + 110, \; left = -4.12, \; right = 4.12 \tag{4.31}$$

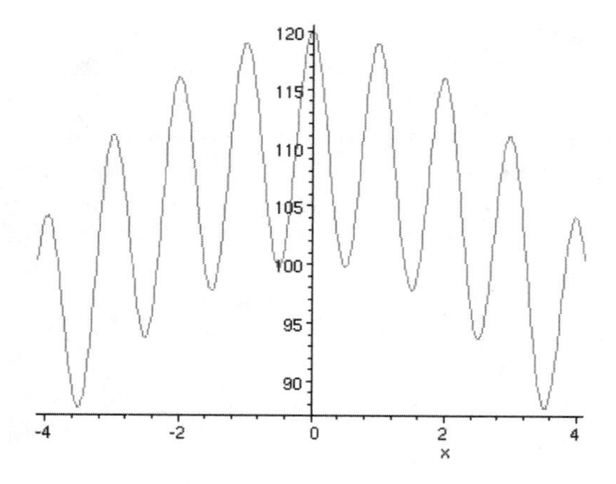

Fig. 4.2. One-dimensional Rastrigin function (see formula 4.31).

which is shown in Figure 4.2.

The next Figure 4.3 presents the chart of the measure associated with the infinite population w defined on $\Omega \cong \{0, 1, \ldots, 255\}$ which is the unique fixed point of the genetic operator in this example. Although the measure is discrete, the chart in the Figure 4.3 has been smoothed to more precisely express its character.

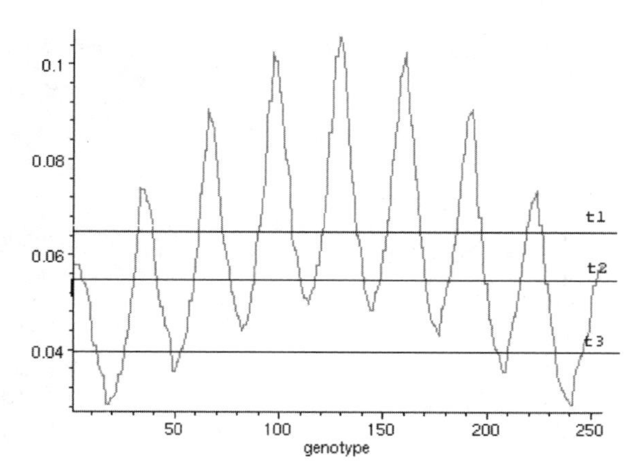

Fig. 4.3. The limit sampling measure associated with the one-dimensional Rastrigin function.

If we normalize the chart of the measure defined on Ω by the total length of $\mathcal{D}$ which equals 8.24 we can deal with the values of the density $\rho_{\theta(w)}$ (see formula 4.28). Figure 4.3 allows us then to discuss the well tuning of the current instance of the SGA to the local maximizers of the objective 4.31.

If *threshold* = t1 then the algorithm is well tuned to all the maximizers excluding the two outermost ones. For *threshold* = t2 the algorithm is well tuned to all the maximizers in $\mathcal{D}$. Finally, if *threshold* = t3 the algorithm is only well tuned to four local maximizers corresponding to the lower local maxims. The basins of attraction of five, central, isolated local maximizers were not distinguished by the level sets of the density function $\rho_{\theta(w)}$.

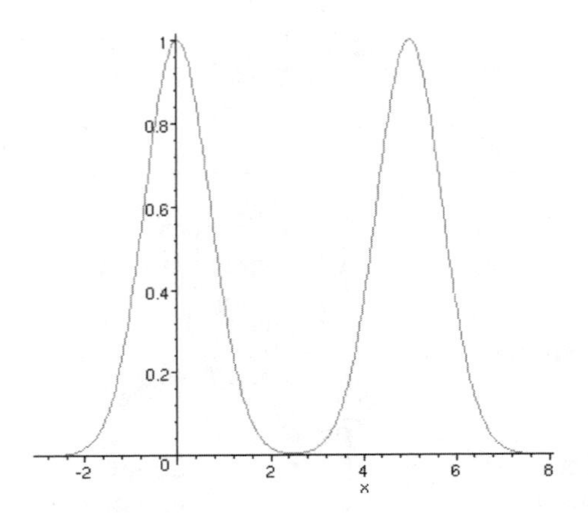

Fig. 4.4. Twin Gauss peaks with different mean values.

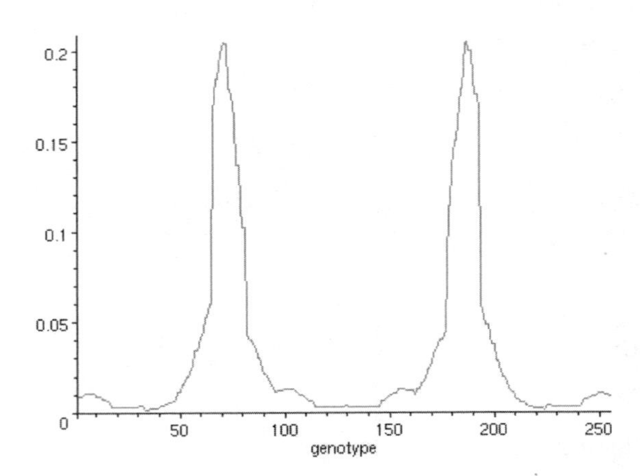

Fig. 4.5. The limit sampling measure associated with twin Gauss peaks with different mean values.

In the next example we will use the objective which is the sum of two Gauss functions which represent the probability distributions with the same standard deviation and different means (see Figure 4.4 for the chart of this function).

$$\Phi_{G1}(x) = \exp(-x^2) + \exp(-(x-5)^2),\ left = -3.81,\ right = 3.81 \quad (4.32)$$

The graph of the limit sampling measure associated with the objective function 4.32 is presented in Figure 4.4. We can easily see that the instance of the Simple Genetic Algorithm is well tuned to both the local maximizers for the wide range of *threshold* parameter values.

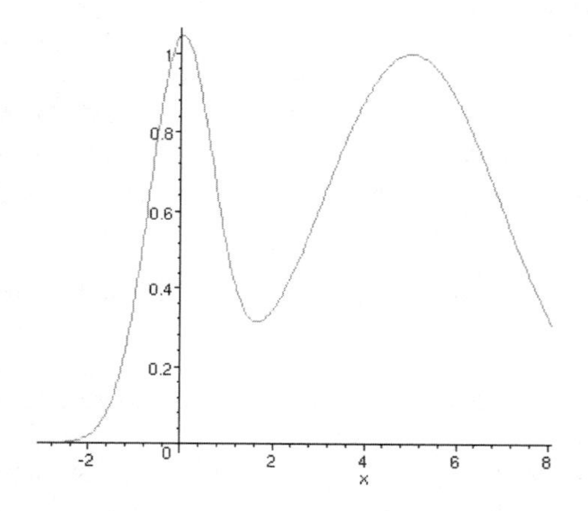

Fig. 4.6. Twin Gauss peaks with different mean and different standard deviation values.

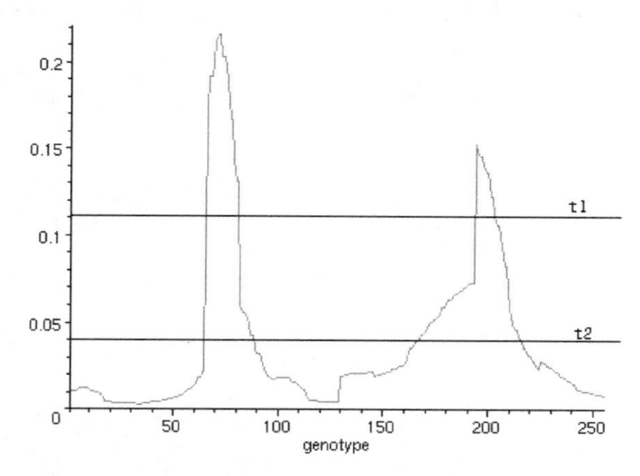

Fig. 4.7. The limit sampling measure associated with twin Gauss peaks with different mean and different standard deviation values.

The last example also deals with the twin Gauss peaks which differ in both: mean and standard deviation (see Figure 4.6).

$$\Phi_{G2}(x) = \exp(-x^2) + \exp(-\frac{1}{8}(x-5)^2),\ left = -3.81,\ right = 3.81 \quad (4.33)$$

The left peak is slender while the right one is wider. The chart of the limit measure is presented in Figure 4.7. As in the previous case we may conclude that the Simple Genetic Algorithm under consideration is well tuned to both local maximizers for the *threshold* parameter values belonging to the wide range included in the range of the measure density variation. Comparing the level sets obtained by *threshold* = t1 and by *threshold* = t2 we may conclude that the sets obtained by $t2$ deliver more information about the shape of the basins of attraction, than obtained by $t1$, so the first, lower *threshold* setting seems to be more suitable for approximating the basins of attraction than the second, larger one.

The examples presented above prove that the well tuning is not an unrealistic postulate, but can be met in many practically motivated algorithm instances. Of course, instances that do not satisfy well tuning condition can also be easily constructed. Special SGA instances in which different fixed points generate limit measures concentrate around different local maximizers (bistability) have recently been presented (see e.g. [205]). The asymptotic behavior of evolutionary algorithm sampling measures with the real number encoding were also studied by Arabas [6].

Let us report now after Schaefer and Jabłoński [154] two main theoretical results of this section.

Theorem 4.65. *Let us assume that the genetic operator* $G : \Lambda^{r-1} \rightarrow \Lambda^{r-1}$ *is focusing and its set of fixed points is finite* ($\#\mathcal{K} < +\infty$) *and the mutation probability is strictly positive* ($p_m > 0$) *then*

$$\forall \varepsilon > 0,\ \forall \eta > 0,\ \exists N \in \mathbb{N},\ \exists W(N) \in \mathbb{N},\ \exists z \in \mathcal{K};$$

so that

$$\forall \mu > N,\ \forall k > W(N),\ \ \forall A \subset \mathcal{D};\ \ A \text{ - measurable in the Lesbeque sense}$$

we have

$$\Pr\left\{\left|\theta'(x_\mu^k)(A) - \theta'(z)(A)\right| < \varepsilon\right\} > 1 - \eta$$

and

$$\Pr\left\{\left|\Psi((\theta(x_\mu^k))(A) - \Psi(\theta(z))(A)\right| < \varepsilon\right\} > 1 - \eta$$

where $x_\mu^k \in \Lambda^{r-1}$ *denotes the frequency vector of the SGA population of size* μ *after* k *genetic epochs.* □

In other words, the counting measure θ' (see formula 4.26 and remark 4.59) of the set A associated with the population of size $\mu < +\infty$ after k genetic epochs bounds with the arbitrarily high probability to the measure associated with the infinite population, which is the fixed point of the genetic operator G, if k and μ are sufficiently large.

Proof. Let us fix $\varepsilon > 0$ and $\eta > 0$. Moreover, the setting $\varepsilon' = c\,\varepsilon$, where c stands for the constant, allows us to bound the Euclidean norm in $\mathbb{R}^r$ by the norm "sum of coordinate modules" from above. All the assumptions of Theorem 4.54 are satisfied, then $\Pr\left\{x_\mu^k \in \mathcal{K}_{\varepsilon'}\right\} > 1 - \eta$ for $\mu > N$ and $k > K(N)$ for the proper sub-class of the SGA spanned by the genetic operator G. It implies, according to the definition of the set $\mathcal{K}_{\varepsilon'}$, that there is $z \in \mathcal{K}$ the fixed point of G so that

$$\Pr\left\{d(x_\mu^k, z) < \varepsilon'\right\} > 1 - \eta.$$

From the equivalence of norms in $\mathbb{R}^r$ we get

$$c \sum_{i=0}^{k-1} \left|(x_\mu^k)_i - z_i\right| < d(x_\mu^k, z)$$

so

$$\Pr\left\{c \sum_{i=0}^{k-1} \left|(x_\mu^k)_i - z_i\right| < \varepsilon'\right\} > 1 - \eta$$

and finally

$$\Pr\left\{\sum_{i=0}^{k-1} \left|(x_\mu^k)_i - z_i\right| < \varepsilon\right\} > 1 - \eta. \tag{4.34}$$

The indices $i = 0, \dots, r-1$ in the above sum run through the whole genotype set Ω and $\mathrm{code}_a(i) \in \mathcal{D}_r \subset \mathcal{D}$. Let us set now $A \subset \mathcal{D}$ and denote by $J \subset \Omega$ the set of genotypes so that $\mathrm{code}_a(i) \in A$ if, and only if, $i \in J$, then

$$\theta(z)(A) = \sum_{i \in J} z_i, \quad \theta(x_\mu^k)(A) = \sum_{i \in J} (x_\mu^k)_i.$$

Now, we may evaluate

$$\left|\theta(z)(A) - \theta(x_\mu^k)(A)\right| = \left|\sum_{i \in J} (z_i - (x_\mu^k)_i)\right|$$

$$\leq \sum_{i \in J} \left|(z_i - (x_\mu^k)_i)\right| \leq \sum_{i=0}^{r-1} \left|(z_i - (x_\mu^k)_i)\right|$$

Taking into account 4.34 we have

$$\Pr\left\{\left|\theta(z)(A) - \theta(x_\mu^k)(A)\right| < \varepsilon\right\} \geq \Pr\left\{\sum_{i=0}^{r-1} \left|(z_i - (x_\mu^k)_i)\right| < \varepsilon\right\} > 1 - \eta$$

which completes the proof of the first thesis of the theorem.

Let us denote now by $J \subset \Omega$ the set of genotypes so that $i \in J$ if, and only if, $\mathrm{meas}(\vartheta_{(i)} \cap A) > 0$, where $\vartheta_{(i)}$ stands for the open brick that contains the phenotype $\mathrm{code}_a(i)$ (see formula 3.14). Because

$$\Psi(\theta(z))(A) = \sum_{i \in J} z_i \frac{\mathrm{meas}(\vartheta_i \cap A)}{\mathrm{meas}(\vartheta_i)},$$

$$\Psi(\theta(x_\mu^k))(A) = \sum_{i \in J} (x_\mu^k)_i \frac{\mathrm{meas}(\vartheta_i \cap A)}{\mathrm{meas}(\vartheta_i)},$$

then

$$\left|\Psi(\theta(z))(A) - \Psi(\theta(x_\mu^k))(A)\right| = \left|\sum_{i \in J} \frac{\mathrm{meas}(\vartheta_i \cap A)}{\mathrm{meas}(\vartheta_i)}(z_i - (x_\mu^k)_i)\right| \le$$

$$\le \sum_{i \in J} \frac{\mathrm{meas}(\vartheta_i \cap A)}{\mathrm{meas}(\vartheta_i)} \left|(z_i - (x_\mu^k)_i)\right| \le \sum_{i=0}^{r-1} \left|(z_i - (x_\mu^k)_i)\right|.$$

Recalling once more formula 4.34 we have

$$\Pr\left\{\left|\Psi(\theta(z))(A) - \Psi(\theta(x_\mu^k))(A)\right| < \varepsilon\right\} \ge \Pr\left\{\sum_{i=0}^{r-1} \left|(z_i - (x_\mu^k)_i)\right| < \varepsilon\right\} > 1 - \eta$$

which completes the proof of the second thesis and then the whole theorem.

□

Theorem 4.66. *Let us assume that the genetic operator* $G : \Lambda^{r-1} \to \Lambda^{r-1}$ *is focusing and its set of fixed points is finite* ($\#\mathcal{K} < +\infty$) *and the mutation probability is strictly positive* ($p_m > 0$) *then*

$$\forall \varepsilon > 0,\ \forall \eta > 0,\ \exists N \in \mathbb{N},\ \exists W(N) \in \mathbb{N},\ \exists z \in \mathcal{K};$$

so that

$$\forall \mu > N,\ \forall k > W(N)\ \ \Pr\left\{\left\|\rho_{\theta(x_\mu^k)} - \rho_{\theta(z)}\right\|_{L^p(\mathcal{D})} < c\,\varepsilon\right\} > 1 - \eta$$

where

$$c = \frac{\mathrm{meas}(\mathcal{D})^{\frac{1}{p}}}{\min_{i \in \Omega}\{\mathrm{meas}(\vartheta_{(i)})\}}$$

and $p \in [1, +\infty)$. □

Proof. Let us select the arbitrary genotype $i \in \Omega$. The thesis of the previous Theorem 4.65 yields

$$\Pr\left\{\left|\Psi(\theta(z))(\vartheta_{(i)}) - \Psi(\theta(x^k_\mu))(\vartheta_{(i)})\right| < \varepsilon\right\} > 1 - \eta \tag{4.35}$$

for $\mu > N$ and $k > W(N)$. Selecting the arbitrary point $\xi \in \text{int}(\vartheta_{(i)})$ we have

$$\Psi(\theta(z))(\vartheta_{(i)}) = \rho_{\theta(z)}(\xi)\,\text{meas}(\vartheta_i)$$
$$\Psi(\theta(x^k_\mu))(\vartheta_{(i)}) = \rho_{\theta(x^k_\mu)}(\xi)\,\text{meas}(\vartheta_i)$$

then from 4.35 we obtain

$$\forall i \in \Omega \quad \Pr\left\{\text{meas}(\vartheta_i)\left|\rho_{\theta(z)}(\xi) - \rho_{\theta(x^k_\mu)}(\xi)\right| < \varepsilon\right\} > 1 - \eta.$$

Because $\text{meas}\left(\mathcal{D} \setminus \bigcup_{i\in\Omega} \vartheta_i\right) = 0$ while $\text{meas}(\mathcal{D}) = \text{meas}\left(\mathcal{D} \cap \bigcup_{i\in\Omega} \vartheta_i\right)$ then for $p \in [1, +\infty)$

$$\Pr\left\{\left\|\rho_{\theta(x^k_\mu)} - \rho_{\theta(z)}\right\|_{L^p(\mathcal{D})} < c\varepsilon\right\} > 1 - \eta$$

where the constant c may be computed as

$$c = \frac{meas(\mathcal{D})^{\frac{1}{p}}}{\min_{i\in\Omega}\{meas(\vartheta_{(i)})\}}$$

which completes the proof. □

Finally, we may draw the concluding remark considering the possibility of solving problem $\mathbf{\Pi_4}$ by the class of SGA spanned by the single genetic operator.

Remark 4.67. If the class of SGA spanned by the genetic operator G is well tuned to the set of local maximizers $\mathcal{W}$ (see definition 4.63) and the assumptions of Theorem 4.66 hold, then the central parts of the basins of attraction $C(x^+), x^+ \in \mathcal{W}$ may be approximated by the level set of the density $\rho_{\theta(x^k_\mu)}$ if μ and k are sufficiently large. □

4.1.3 The results of the Markov theory for Evolutionary Algorithm

The asymptotic behavior of evolutionary algorithms of the type $(\mu + \lambda)$ (see 3.9) will be analyzed in this section. We will utilize problem $\mathbf{\Pi_1}$ (find any global maximizer only, see Section 2.1) without constraints, i.e. $\mathcal{D} = V$. We will apply the phenotypic encoding for which the genetic universum

$$U = \mathbb{R}^N \tag{4.36}$$

may also be identified with the search space V. The space of state for this group of algorithms is

$$\mathcal{E} = U^\mu/eqp = (\mathbb{R}^N)^\mu/eqp \tag{4.37}$$

where the equivalence is described by the formula 2.14 and μ stands for the constant (does not depend on the number of the genetic epoch) population cardinality.

We will use the elitist selection (see Section 3.4.3) which transports the best fitted individual to the next epoch from the current population with the probability 1, and mutation described by the simplified version of the formula 3.51. We will also apply other phenotypic genetic operations (e.g. crossover described by the formula 3.52) which do not memorize any population characteristics.

Because neither genetic operation which is applied to create the next epoch population P_{t+1} depends on the number of the genetic epoch t as well as on the previous populations $P_0, P_1, \ldots, P_t$, then the evolutionary algorithm under consideration may be modeled as the uniform Markov chain with the space of states $\mathcal{E}$ (see formula 4.37) and the Markov transition function

$$\tau : \mathcal{E} \rightarrow \mathcal{M}(\mathcal{E}). \tag{4.38}$$

Please, note that the space of states is uncountable $\#\mathcal{E} > \#\mathbb{N}$ in this case, even if the population cardinality is finite $\mu < +\infty$.

Almost all the results presented in this section are cited from papers written by Rudolph [140], [141], [143], [144], Beyer and Rudolph [27] and Grygiel [82].

The model of the $(\mu + \lambda)$ algorithm with elitist selection

We start with the analysis of the very simple $(1+1)$ evolutionary algorithm with a single-individual population $P_t = \{x_t\},\ t \geq 0$. The single individual is only mutated in each genetic epoch by adding the random vector

$$z_t = \mathcal{N}(0, \sigma\mathbf{I}) \tag{4.39}$$

where $\mathcal{N}(0, \sigma\mathbf{I})$ denotes the N-dimensional random variable with a normal probability distribution, $\sigma > 0$ stands for the mean standard deviation and $\mathbf{I}$ denotes the $N \times N$ diagonal identity matrix. Moreover, we assume that z_k is independent of z_l for $k \neq l$.

The mutated individual $y_t = x_t + z_t$ is selected for the next epoch population P_{t+1} if $f(x_t) < f(y_t)$. If not, P_{t+1} is set as P_t. Such a selection is called a hard elitist one (see Section 3.4.3).

The space of states will be $\mathcal{E} = U = \mathcal{D} = \mathbb{R}^N$ in this case. The Markovian kernel of such mutation will be expressed by the formula

$$\tau_m(x)(A) = \int_A \rho_{\mathcal{N}(0,\sigma\mathbf{I})}(z - x)\, dz \tag{4.40}$$

where $x \in \mathcal{E}, A \subset \mathcal{E}$ is the measurable set in the Lesbegue sense, $\rho_{\mathcal{N}(0,\sigma\mathbf{I})}$ stands for the density function of the probability distribution of $\mathcal{N}(0, \sigma\mathbf{I})$.

Let us define now

$$W(x) = \{y \in \mathcal{E};\ f(y) > f(x)\} \tag{4.41}$$

which is the set of admissible solutions which are not worse fitted than $x \in \mathcal{E}$. Because the result of selection depends on the previous state $x \in \mathcal{E}$, then the Markovian kernel of selection depends on this quantity

$$\tau_s(y,x)(A) = \chi_{W(x)}(y)\chi_A(y) + \chi_{(W(x))^-}(y)\chi_A(x) \tag{4.42}$$

where $(W(x))^- = \mathcal{E} \setminus W(x)$ denotes the complement of $W(x)$ and χ_A is the characteristic function of the set A.

Let us note that:

- If $y \in \mathcal{E}$ is better or equally fitted than $x \in \mathcal{E}$ (i.e. $y \in W(x)$) then the passage from x to $y \in A$ (or more correctly to $y \in A \cap W(x)$) occurs with the probability 1.
- If $y \in \mathcal{E}$ is worse fitted than x (i.e. $y \in (W(x))^-$) then y is not accepted. $\tau(y,x)(A) = 1$ only if $x \in A$ in this case.
- All other cases have the occurrence probability 0.

The selection kernel is deterministic as is usual for hard selection. In order to obtain the Markovian kernel of the discussed algorithm both kernels τ_m and τ_s should be composed (see remark 4.10).

$$\begin{aligned}
\tau(x)(A) &= \int_{\mathcal{E}} \tau_m(x)(dy) \cdot \tau_s(y,x)(A) = \\
&= \left(\int_{\mathcal{E}} \tau_m(x)(dy)\chi_{A\cap W(x)}(y)\right) + \chi_A(x)\int_{\mathcal{E}} \tau_m(x)(dy)\chi_{(W(x))^-}(y) = \\
&= \left(\int_{A\cap W(x)} \tau_m(x)(dy)\right) + \chi_A(x)\int_{(W(x))^-} \tau_m(x)(dy) = \\
&= \tau_m(x, A\cap W(x)) + \chi_A(x)\tau_m(x)((W(x))^-)
\end{aligned} \tag{4.43}$$

Now we try to extend the above modeling result to the case of populations containing more than one individual ($\mu > 1$). We start from two very simple observations:

Remark 4.68. Both formulas 4.42, 4.43 which were introduced for the $(1+1)$ algorithm for which $\mathcal{E} = U = \mathbb{R}^N$ can be extended to the arbitrary space of states $\mathcal{E}$ if the rule of passing a "better" fitted population to the next epoch is reconstructed. □

Remark 4.69. The kernel operator τ_m defined by formula 4.40 for the $(1+1)$ algorithm may be replaced by the system of mixing operations that modify the state $x \in \mathcal{E}$ in the stochastic manner. □

Let us consider now the genetic universum $U = \mathcal{D}$ and the space of states $\mathcal{E} = U^{\mu}/eqp$, $\mu < +\infty$. We may introduce the mapping

$$b : \mathcal{E} \ni x \to b(x) \in U \tag{4.44}$$

which selects the genotype of the best fitted individual $b(x)$ in the population $x \in \mathcal{E}$. Similarly to the previous rule we may select the set of states which are "not worse" than the particular state $x \in \mathcal{E}$ such as

$$W(x) = \{x \in \mathcal{E};\ f(b(y)) \geq f(b(x))\} \tag{4.45}$$

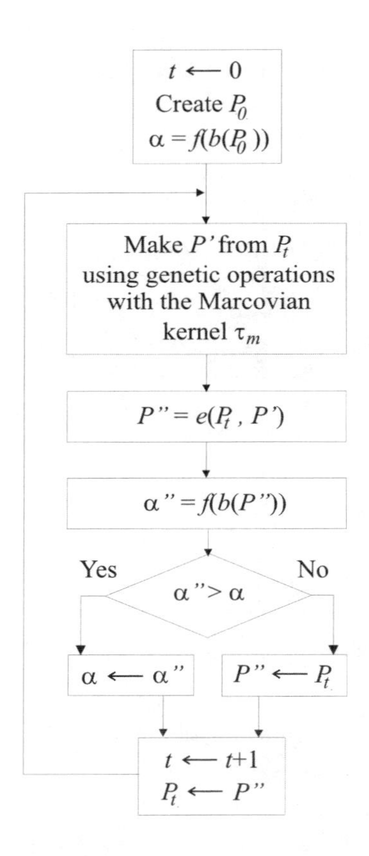

Fig. 4.8. The scheme of the evolutionary algorithm with elitist selection on the population level and deterministic passage of the best fitted individual.

Now we will utilize the selection rule which is applied immediately to the whole population $y \in \mathcal{E}$ which can be obtained from the previous epoch population $x \in \mathcal{E}$:

1. If $y \in W(x) \cap A$, then the algorithm leaves the set of states A and the next population is $y \in \mathcal{E}$.
2. If $y \notin W(x)$ then the best fitted individual $b(y)$ is "worse" than $b(x)$ and the whole population y is rejected. The population $x \in \mathcal{E}$ passes to the next epoch.

If the above selection rule is applied, then the formula 4.42 which determines τ_m remains valid. Moreover, we need one new mapping

$$e : \mathcal{E} \times \mathcal{E} \ni (x, y) \to y' = e(x, y) \in \mathcal{E} \tag{4.46}$$

where the population y' is obtained from the population y by placing in y the individual $b(x)$ in some way. The scheme of such algorithms of the type $(\mu + \lambda)$ is presented in Figure 4.8. Now the selection may be characterized by the following Markovian kernel

$$\tau_s(x, y)(A) = \chi_{A \cap W(x)}(y) + \chi_{(W(x))^-}\chi_A(x)\chi_A(e(x, y)). \tag{4.47}$$

Composing now τ_s with the mixing kernel τ_m we obtain

$$\tau_A(x)(A) = \tau_m(x)(W(x) \cap A) + \chi_A(x) \int_{(W(x))^-} \tau_m(x)(dy)\chi_A(e(x, y)) \tag{4.48}$$

where the above integral has to be computed according to the proper measure on the space of states $\mathcal{E}$.

The convergence of Evolutionary Algorithm to the global maximizer

Let us consider again global optimization problem $\mathbf{\Pi_1}$ (see Section 2.1) which consists of finding at least one global maximizer for the objective function Φ on the admissible set $\mathcal{D}$. The evolutionary algorithm of the $(\mu + \lambda)$ type with the genetic universum $U = \mathcal{D}$ and the fitness function $f = \Phi$ will be utilized for solving $\mathbf{\Pi_1}$.

If $x^* \in \mathcal{D}$ is the solution to $\mathbf{\Pi_1}$ then $f^* = f(\text{code}(x^*))$ will denote the maximum fitness value that may appear in this instance of evolutionary computation. Let $P_1, P2, \ldots$ be the stochastic sequence of populations produced by this algorithm. We define the new stochastic sequence (stochastic process)

$$\{Y_t = f(b(P_t))\},\ t \geq 0 \tag{4.49}$$

which is instantiated by the best fitted individuals in the consecutive populations. Such a sequence may be considered as the sequence of estimators of f^*.

It is interesting to know when the sequence $\{Y_t\}$, $t \geq 0$ converges to f^* and what sort of convergence this is. Let us introduce the family of level sets

$$A_\varepsilon = \{x \in \mathcal{E};\ (f^* - f(b(x))) \leq \varepsilon\},\ \varepsilon > 0 \tag{4.50}$$

The answer to the above question is delivered by the following theorem.

Theorem 4.70. *(see Rudolph [141]) Assume that the fitness function* $f : U \to \mathbb{R}_+$ *satisfies*

$$\exists M,\ 0 < M < +\infty;\ f(x) \leq M\ \forall x \in U$$

Let us consider the evolutionary algorithm with the space of states $\mathcal{E} = U^\mu/eqp$ *whose dynamic may be described by the Markovian kernel* τ *that satisfies following conditions:*

$$\forall \varepsilon > 0\ \exists \delta > 0 \ \textit{such that}$$

$$\tau(x)(A_\varepsilon) \geq \delta\ \ \forall x \in (A_\varepsilon)^-,$$

$$\tau(x)(A_\varepsilon) = 1\ \ \forall x \in A_\varepsilon,$$

then the sequence $\{Y_t\}$, $t \geq 0$ *converges completely to* f^* *which means that*

$$\forall \varepsilon > 0\ \ \left\{ \lim_{t\to+\infty} \sum_{i=0}^{t} \Pr\{(f^* - Y_i) > \varepsilon\} \right\}, +\infty$$

□

Remark 4.71. If the sequence $\{Y_t\}$, $t \geq 0$ converges completely to f^* then it also almost surely converges to f^* i.e.

$$\Pr\left\{ \lim_{t\to+\infty} (f^* - Y_t) = 0 \right\} = 1$$

and converges in probability to f^* i.e.

$$\forall \varepsilon > 0\ \ \lim_{t\to+\infty} \Pr\{(f^* - Y_t) > \varepsilon\} = 0.$$

□

Theses of the above remark follow the well-known dependencies between the various sorts of stochastic convergence that may be found in many monographs (see e.g. Lucas [106], Chow and Teicher [49], Iosifescu [88], Billingsley [28]).

Example 4.72. Let us check when the group of algorithms described in the previous sections satisfy the assumptions of Theorem 4.70. The Markovian kernel of such algorithms is described by the formula 4.48. If $A_\varepsilon \subset W(x)$ and $A_\varepsilon \neq W(x)$, then for $x \notin A_\varepsilon$ we have $A_\varepsilon \cap W(x) = A_\varepsilon$ where $W(x) = \{y \in \mathcal{E};\ f(b(y)) \geq f(b(x))\}$ and then

$$\tau(x)(A_\varepsilon) = \tau_m(x)(A_\varepsilon).$$

If now $W(x) \subseteq A_\varepsilon, x \in A_\varepsilon$ and $W(x) \cap A_\varepsilon = W(x)$ then we have

$$\tau(x)(A_\varepsilon) = \tau_m(x)(W(x)) + \int_{(W(x))^-} \tau_m(x)(dy)\ \chi_{A_\varepsilon}(e(x,y)) =$$
$$= \tau_m(x)(W(x)) + \tau_m(x)((W(x))^-) = 1$$

because $e(x,y) \in W(x) \subseteq A_\varepsilon$. The second assumption of Theorem 4.70 is then satisfied.

The probability of the passage to the set A_ε can be computed as

$$\tau_m(x)(A_\varepsilon) = \tau_m(x)(A_\varepsilon)\,\chi_{(A_\varepsilon)^-}(x) + \chi_{A_\varepsilon}(x).$$

The first assumption of Theorem 4.70 will be satisfied if $\tau_m(x)(A_\varepsilon) \geq \delta,\ \forall x \in (A_\varepsilon)^-$ for some $\delta > 0$.

Now we intend to formulate the sufficient conditions for the convergence of the discussed class of algorithms. Let us assume that the mixing operation is composed of mutation and other operations (e.g. crossover). The mixing kernel τ_m will be the composition of the mutation kernel τ_{mut} and the kernel τ_c that models the remaining mixing operations. Moreover, we assume that the space of states of the considered class of algorithms is finite i.e. $\#\mathcal{E} < +\infty$.

The crucial, practical assumption will be stressed as follows: mutation allows us to obtain the arbitrary state $x' \in \mathcal{E}$ in the single genetic epoch starting from another arbitrary state $x \in \mathcal{E}$. In other words

$$\exists \sigma_m > 0;\ \ \tau_{mut}(x)(\{x'\}) \geq \sigma_m,\ \ \forall x, x' \in \mathcal{E}. \tag{4.51}$$

Then it implies

$$\begin{aligned}\tau_{mut}(x)(\{x'\}) &= \sum_{y\in\mathcal{E}} \tau_c(x)(\{y\})\,\tau_{mut}(y)(\{x'\}) \geq \\ &\geq \sigma_m \sum_{y\in\mathcal{E}} \tau_c(x)(\{y\}) = \sigma_m \tau_c(x)(\mathcal{E}) = \sigma_m > 0\end{aligned} \tag{4.52}$$

and also $\tau_m(x)(A_\varepsilon) \geq \sigma_m > 0\ \ \forall x \in (A_\varepsilon)^-$, moreover, $A_\varepsilon \neq \emptyset$.

The above result may be compared to Theorem 4.31 and the remark 4.32 whih is the consequence of applying the ergodic theorem to the Markov chain that models the Simple Genetic Algorithm. □

The convergence of Supermartingals in Evolutionary Algorithm convergence analysis

Definition 4.73. *(see Neveu [114]) Let the triple* $(\Omega_P, \Im, \Pr)$ *be the probabilistic space where* Ω_P *stands for the set of elementary events,* $\Im$ *is the* σ*-algebra and* Pr *the probability measure. We will analyze the increasing sequence of sub-*σ*-algebras* $\Im_0 \subseteq \Im_1 \subseteq \Im_2 \subseteq \cdots \subseteq \Im$ *that satisfies the condition*

$$\Im_\infty = \sigma\left(\bigcup_t \Im_t\right) \subseteq \Im$$

where $\sigma(\mathcal{A})$ *denotes the minimum* σ*-algebra that contains the class* $\mathcal{A}$ *of subsets of the elementary event set* Ω_P *(see Billingsley [28]). The stochastic process* $\{X_t\}$, $t \geq 0$ *for which random variables* X_t *are* $\Im_t$*-measurable will be called supermartingal if*

$$\forall t \geq 0 \;\; \mathrm{E}(|X_t|) < +\infty \;\; \textit{and} \;\; \mathrm{E}(X_{t+1}\,|\Im_t) \leq X_t \;\textit{almost surely}$$

where $\mathrm{E}(\cdot\,|\Im)$ *denotes the expected value operator computed with respect to the* σ*-algebra* $\Im$ *(see e.g. Billingsley [28]). Moreover, if*

$$\forall t \geq 0 \;\; \Pr\{X_t \geq 0\} = 1$$

then the supermartingal $\{X_t\}$, $t \geq 0$ *will be called non-negative.* □

Theorem 4.74. *(see Rudolph [140]) Let* $\{X_t\}$, $t \geq 0$ *be the non-negative supermartingal that satisfies*

$$\mathrm{E}(X_{t+1}\,|\Im_t) \leq c_t\, X_t \;\textit{with the probability 1}$$

$$\textit{where}\;\; c_t \geq 0 \;\textit{for}\; t \geq 0 \;\textit{and}\; \sum_{t=1}^{+\infty}\left(\prod_{k=0}^{t-1} c_k\right) < +\infty$$

then

$$\lim_{t\to+\infty} \mathrm{E}(X_t) = 0$$

which means that $\{X_t\}$, $t \geq 0$ *converges in mean and converges completely*

$$\forall \varepsilon > 0 \;\; \left(\lim_{t\to+\infty} \sum_{i=0}^{t} \Pr\{X_i > \varepsilon\}\right) < +\infty.$$

□

Remark 4.75. The thesis of the theorem 4.74 also implies the almost surely convergence of the supermartingal $\{X_t\}$, $t \geq 0$ to zero

$$\Pr\left\{\lim_{t\to+\infty} |X_t| = 0\right\} = 1$$

and convergence in probability

$$\forall \varepsilon > 0 \;\; \lim_{t\to+\infty} \Pr\{|X_t| > \varepsilon\} = 0.$$

□

The above remark is the simple issue from well-known dependencies among various modes of stochastic convergence (see e.g. Lucas [106], Chow and Teicher [49]). We will continue with two almost trivial observations.

Remark 4.76. The assumptions of the above Theorem 4.74 are satisfied if

$$\limsup_{t \geq 0} \{c_t\} < 1$$

or in particular $c_t \equiv c < 1$. □

Remark 4.77. If $c_t \equiv c < 1$ then for $t \geq 0$ the inequality

$$E(X_{t+1} \,|\Im_t\,) \leq c\, X_t$$

holds almost surely, then

$$E(X_{t+1} \,|\Im_t\,) \leq c\, E(X_t)$$

and

$$E(X_t) \leq c^t\, E(X_0)$$

so the rate of convergence of the supermartingal $\{X_t\},\ t \geq 0$ to zero is geometrical. □

Taking the above theorem and remarks into account, we try to formulate the other necessary conditions for the evolutionary algorithm convergence. These conditions will handle the expected increment of the maximum fitness $f(b(P_t))$ that occurs in the population in the particular evolution step.

Let us assign to the evolutionary algorithm that processes populations $P_t \in \mathcal{E} = U^\mu / eqp$ a new random sequence

$$\{\omega_t\} = \{f^* - f(b(P_t))\},\ t \geq 0 \tag{4.53}$$

where $f : U \rightarrow \mathbb{R}_+$ stands for the fitness function and f^* for its maximum value.

Remark 4.78. If $f < M < +\infty$ on the whole U and the sequence of populations $\{P_t\},\ t \geq 0$ was produced by the evolutionary algorithm described in the example 4.72 then the random sequence $\{\omega_t\},\ t \geq 0$ is the non-negative suprmartingal. □

Proof. The expectation $E(|\omega_t|) = E\,(|f^* - F(b(P_t))|)$ takes finite values when the fitness function f is bounded on U. The random variable ω_t is non-negative almost surely for $t \geq 0$ because $f^* \geq f(b(P_t)),\ t \geq 0$. Moreover, $E(\omega_{t+1} \,|\Im_t\,) \leq \omega_t$ holds almost surely because $f(b(P_{t+1})) \geq f(b(P_t)),\ t \geq 0$. □

Remark 4.79. Let the evolutionary algorithm described in example 4.72 satisfy the assumptions of remark 4.78 and there is the constant $c \in (0,1)$ so that

$$E\,(f^* - f(b(P_{t+1}))\,|\Im_t\,) \geq c\,(f^* - F(b(P_t))$$

then the sequence $\{f_t\} = \{f(b(P_t))\},\ t \geq 0$ converges in mean to f^* i.e.

$$\lim_{t \rightarrow +\infty} E(|f^* - f(b(P_t))|) = 0$$

with the geometrical rate c^t which precisely means that $|f^* - f(b(P_t))|$ has the order $O(c^t)$. □

Proof. Of course, if the assumptions of remarks 4.78 and 4.79 are satisfied then the random sequence $\{\omega_t\}$, $t \geq 0$ is the non-negative supermartingal which satisfies all the assumptions of Theorem 4.74, so according to its thesis the in mean convergence may be drawn $\lim_{t \to +\infty} \mathrm{E}(\omega_t) = 0$ and then

$$\lim_{t \to +\infty} \mathrm{E}(f^* - f(b(P_t))) = f^* - \lim_{t \to +\infty} \mathrm{E}(f(b(P_t))) = 0.$$

The geometrical rate of such convergence may be drawn from remark 4.77. □

4.2 Asymptotic results for very small populations

Some interesting asymptotic results have been obtained for the genetic algorithms which process very small (1 - 2 individuals) populations. Although they do not deliver tools for immediate analysis of computation instances applied in engineering practice, they allow us to establish the intuition necessary for better understanding of the particular genetic operation's activity and synergy. They are also helpful when trying to understand the meaning of various convergence modes associated with genetic algorithm dynamics.

The first group of results described in Section 4.2.1 was formulated and proved by Kazimierz Grygiel as well as published in his paper [81]. The second one stressed in Section 4.2.2 was delivered mainly by Iwona Karcz-Dulęba and published in the papers [70], [62] and [91].

4.2.1 The rate of convergence of the single individual population with hard succession

We intend to evaluate the rate of convergence of the genetic algorithm that solves one-dimensional optimization problem of type $\mathbf{\Pi_1}$ (see Section 2.1)

$$\max_{x \in S}\{\Phi(x)\}, \quad S = [0,1] \subset \mathbb{R} \tag{4.54}$$

for the unimodal, continuous objective function $\Phi :\to \mathbb{R}_+$. Besides the simplicity of the problem to be solved, the genetic computation model was also drastically simplified in order to obtain a strong mathematical result.

We will apply the particular type of the affine binary encoding in which we utilize the binary strings of the length l as genotypes. The binary universum $\Omega = \{a = (a_0, \dots, a_{l-1});\ a_j \in \{0,1\},\ 0 \leq j \leq l-1\}$ will be mapped on the left ends of the sub-intervals of the length 2^{-l}, so the encoding and the set of phenotypes will be defined by the formula

$$\begin{aligned} \mathrm{code} : \Omega \ni a \to \mathrm{code}(a) = \textstyle\sum_{j=0}^{l-1} a_j 2^{-j-1} \in \mathcal{D}_l \\ \mathcal{D}_l = \{\mathrm{code}(a), a \in \Omega\}. \end{aligned} \tag{4.55}$$

Please note that, in contrast to the traditional binary affine encoding, the right end of the interval S does not belong to the phenotypes $1 \notin \mathcal{D}_l$. Because the numerical value of each genotype $a \in \Omega$ precisely fits the phenotype code(a), then we will not distinguish among the genotype, and associated phenotypes, if it does not lead to the ambiguity.

We will use the single genetic operation called AB-mutation, which is given by the stochastic mapping

$$\Omega \ni a \to a \oplus \pm i \in \Omega. \tag{4.56}$$

The code $i \in \Omega$ stands for the AB-mutation mask which is obtained by sampling according to the same probability distribution as in the case of the binary multi-point mutation (see Section 3.5.1, formula 3.37).

$$\Pr(\{i\}) = (p_m)^{(\mathbf{1},i)}(1-p_m)^{l-(\mathbf{1},i)} \tag{4.57}$$

where

$$\mathbf{1} = \underbrace{(1,\ldots,1)}_{l-times}$$

denotes the binary vector of l units and $(\mathbf{1}, i)$ the Euclidean scalar product of binary vectors $\mathbf{1}$ and i, moreover, $p_m \in [0,1]$ stands for the mutation rate parameter.

The signum which is assigned to the mask is also sampled from the set $\{+,-\}$ with the uniform probability distribution $\{\frac{1}{2},\frac{1}{2}\}$ independent of the mask code i.

The suggested type of mutation exhibits a much stronger ability of the interval S exploration than the standard multi-point binary mutation, which replaces the individual $a \in \Omega$ by $a \oplus i$. It may be formalized as the following lemma.

Lemma 4.80. *(see Grygiel [81], Lemma 1) Let $x, y \in \mathcal{D}_l,\ x < y$ then there are two masks $i', i'' \in \Omega;\ (\mathbf{1}, i') = (\mathbf{1}, i'') = 1$ so that*

$$y - x' < \frac{y-x}{2} \quad \text{and} \quad x'' - x < \frac{y-x}{2}$$

where $x' = x + i'$ and $x'' = x + i''$. □

In other words, for each of the two nodes x, y there is a one-point mutation mask with a large sampling probability (see formula 4.57) which produces the individual that is less distant to the parent than to the second node.

The genetic algorithm under consideration transforms the single individual populations, then $\mathcal{E} = \mathcal{D}_l$ stands for its space of states. The transition to the next state is described by the formula

$$x_{t+1} = \begin{cases} \xi_t & \text{if } \Phi(\xi_t) > \Phi(x_t) \\ x_t & \text{in the other case} \end{cases}, \quad t \geq 0 \tag{4.58}$$

where ξ_t is the AB-mutant of x_t. The above algorithm may be understood as the instance of the random walk (see Section 2.2) in which AB-mutation plays the role of the sampling procedure. Let us denote by $\hat{\Phi} = \Phi\,|\mathcal{D}_l$ the restriction of the objective function to the phenotype set. The main result of this section will be formulated as follows.

Theorem 4.81. *(see Grygiel [81], Theorem 1) For the arbitrary unimodal, bounded function $\hat{\Phi} : \mathcal{D}_l \to \mathbb{R}_+$ the expected time T_l to find the best approximation of the global maximizer satisfies the inequality*

$$T_l \leq \frac{4\,l}{p_m(1-p_m)^{l-1}}$$

□

Remark 4.82. The denominator in the right hand side of the main formula in Theorem 4.81 reaches the maximum value if $p_m = \frac{1}{l}$ which is the best hint for choosing algorithm parameters $(l, p_m = \frac{1}{l})$. Moreover, for such parameter setting $T_l = O(l^2)$, because $\left(1 - \frac{1}{l}\right)^{l-1} \to e^{-1}$ for $l \to +\infty$. □

The above remark may be generalized to the case of multidimensional, bounded, unimodal functions $\hat{\Phi} : (\mathcal{D}_l)^N \to \mathbb{R}_+$ which are separable, i.e. there exists a representation of $\hat{\Phi}$, so that

$$\hat{\Phi}(x_1, \ldots, x_N) = \sum_{i=1}^{N} \alpha_i \hat{\Phi}_i(x_i),\ \alpha_i \neq 0,\ \hat{\Phi}_i : \mathcal{D}_l \to \mathbb{R}_+,\ i = 1, \ldots, N. \quad (4.59)$$

In this case, the optimal parameter selection is $(l, p_m = \frac{1}{N\,l})$ and gives the evaluation $T_l = O(N\,l^2)$.

4.2.2 The dynamics of double individual populations with proportional selection

This section presents results related to the dynamics of the populations that are composed of only two individuals whose genotypes are real-valued vectors. Similarly to the previous Section 4.2.1, mutation is the only genetic operation that will be used by the new population creation. However, asymptotic results will be of a different sort and will be quite differently presented. The section contents are based on the papers [70], [62] and [91].

The presentation will start with a detailed description of the slightly more general case of evolutionary algorithms for which each population is of the constant size $\mu \in \mathbb{N}$, not only for $\mu = 2$. The genetic universum will be $U = \mathbb{R}^N$ (see phenotypic encoding, Section 3.1.3). The space of states will then be $\mathcal{E} = (\mathbb{R}^N)^\mu / eqp$, where eqp is the equivalence defined in Section 2.2 (see formula 2.14) which identifies the μ-element strings of N-dimensional vectors, which may be obtained by the permutation of their vector elements.

The global optimization problem under consideration consists of finding the global maximizer to the real-valued, bounded function of N real variables in the unbounded domain - the unbounded domain version of problem $\mathbf{\Pi_1}$ defined in Section 2.1.

$$\max_{x \in \mathbb{R}^N} \{f(x)\}, \ f(x) < M < +\infty, \ \forall x \in \mathbb{R}^N \tag{4.60}$$

The stochastic transition rule $\tau : \mathcal{E} \to \mathcal{M}(\mathcal{E})$ may be defined in the following steps:

1. Select the individual x from the population P according to the proportional selection rule i.e. x will be obtained by multiple sampling according to the probability distribution given by formula 3.30 (see Section 3.4.1).
2. Modify the individual $x \to x'$ according to the phenotypic, normal mutation rule (see Section 3.7.1, formula 3.51), i.e. $x' = x + \xi$ where ξ is the result of sampling according to the N-dimensional normal distribution with a zero mean and the diagonal, isotropic covariance matrix $\mathbf{C} = \sigma \mathbf{I}$. The coefficient σ stands for the standard deviation of this distribution which is valid for each dimension in the genetic universum.
3. Place the offspring x' in the next epoch population P'.
4. If the next epoch population P' contains less than μ individuals, then go to step 1.

The probability of sampling the individual $x \in U = \mathbb{R}^N$ to the next epoch population P' is given by the formula

$$\Pr(\{x\}|P) = \sum_{y \in P} \alpha(y) \rho_{\mathcal{N}(y,\sigma\mathbf{I})}(x) \tag{4.61}$$

where

$$\alpha(y) = \frac{f(y)}{\sum_{z \in P} f(z)}$$

is the probability of selecting the individual with the genotype y from the current population P and $\rho_{\mathcal{N}(y,\sigma\mathbf{I})}$ is the density of the N-dimensional normal distribution with the mean $y \in U$ and the covariance matrix $\mathbf{C} = \sigma \mathbf{I}$.

The probability distribution $\tau(P) \in \mathcal{E}$ has the density function. Let $P' = \langle x_1, \ldots x_\mu \rangle \in \mathcal{E}$ be the population that immediately follows P, then

$$\tau(P)(\{P'\}) = \mu! \prod_{j=1}^{\mu} \Pr(\{x_j\}|P) = \mu! \prod_{j=1}^{\mu} \sum_{y \in P} \alpha(y) \rho_{\mathcal{N}(y,\sigma\mathbf{I})}(x). \tag{4.62}$$

If $\mu = 2, \ N = 1$ (two individual population, scalar genotypes) then we may represent each population as $P = \{y_1, y_2\}, y_i \in \mathbb{R}, \ i = 1, 2$. The probability of sampling the individual with the genotype x to the epoch that follows P is given by the formula

$$\Pr(\{x\}|P) = \alpha(y_1) \rho_{\mathcal{N}(y_1,\sigma)}(x) + \alpha(y_2) \rho_{\mathcal{N}(y_2,\sigma)}(x) \tag{4.63}$$

where $\rho_{\mathcal{N}(y,\sigma)}(x)$ stands now for the density function of the unidimensional normal probability distribution with the mean value y and the standard deviation σ. For the next epoch population $P' = \{x_1, x_2\}$ we have

$$\begin{aligned}\tau(P)(\{P'\}) =& 2\ \Pr(\{x_1\}|P)\ \Pr(\{x_2\}|P) = \\ =& 2\left(\alpha(y_1)\rho_{\mathcal{N}(y_1,\sigma)}(x_1) + \alpha(y_2)\rho_{\mathcal{N}(y_2,\sigma)}(x_1)\right) \\ & \left(\alpha(y_1)\rho_{\mathcal{N}(y_1,\sigma)}(x_2) + \alpha(y_2)\rho_{\mathcal{N}(y_2,\sigma)}(x_2)\right).\end{aligned} \tag{4.64}$$

The space of states of the evolutionary algorithm may be interpreted now as the half-plane according to the following rule

$$\mathcal{E} \ni (x_1, x_2) \rightarrow \begin{cases} (x_1, x_2) \ \ if \ \ x_1 \geq x_2 \\ (x_2, x_1) \ \ if \ \ x_1 < x_2. \end{cases} \tag{4.65}$$

This is possible because the state of the algorithm does not depend on the order of the individuals in the population. The next formal step that makes the analysis more convenient is the $\frac{\pi}{4}$ turn of the system of coordinates (x_1, x_2) in the space of states. The new system of coordinates (w, z) will be given by the formula

$$w = \frac{(x_1 - x_2)}{\sqrt{2}},\ z = \frac{(x_1 + x_2)}{\sqrt{2}},\ w \geq 0. \tag{4.66}$$

In this system of coordinates the expected location (w', z') of the next epoch population P' is computed with respect to the location of the current population P located at (w, z).

$$\begin{gathered}\mathrm{E}(w'|P) = \sigma\sqrt{\frac{2}{\pi}} + (1-\gamma^2)\,\sigma\left(\phi\left(\frac{w}{\sigma}\right) + \frac{w}{\sigma}\,\Xi\left(\frac{w}{\sigma}\right)\right) \\ \mathrm{E}(z'|P) = z + \gamma w,\ \ \gamma = \frac{q_1 - q_2}{q_1 + q_2} \\ q_1 = f(x_1) = f\left(\frac{w+z}{\sqrt{2}}\right),\ \ q_2 = f(x_2) = f\left(\frac{z-w}{\sqrt{2}}\right) \\ \phi(\zeta) = \frac{1}{\sqrt{2\pi}}\left(\exp\left(-\frac{\zeta^2}{2}\right) - 1\right),\ \ \Xi(\zeta) = \frac{1}{\sqrt{2\pi}}\int_0^{\zeta}\exp\left(-\frac{t^2}{2}\right)dt\end{gathered} \tag{4.67}$$

Next computations are performed for two fitness functions

$$\begin{gathered} f(x) = \exp(-ax^2), \\ f(x) = \exp(-a(x+d)^2) + \exp(-a(x-d)^2) \end{gathered} \tag{4.68}$$

where a and d are the positive, numerical parameters. Let us now compute the expected locations for populations of some characteristic arguments (previous states, previous step populations), which allow us to formulate substantial qualitative conclusions according to the double individual population dynamic

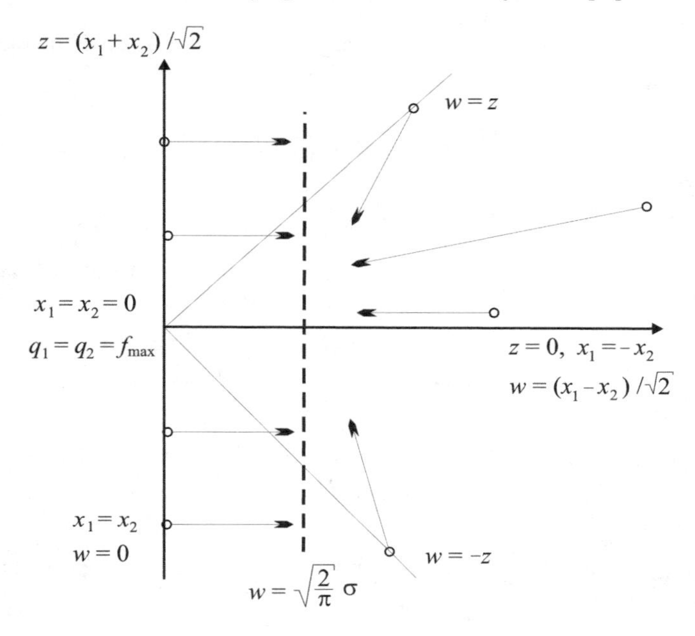

Fig. 4.9. Expected behavior of the two-individual population.

behavior. Such expected locations have a similar meaning to the values of the genetic operator (see definition 4.16 and Theorem 4.20) for the Simple Genetic Algorithm.

1. If the population P lies on the $0z$ axis ($w = 0$), i.e. $x_1 = x_2,\ q_1 = q_2$ then

$$\mathrm{E}(w'|P) = \sigma\sqrt{\frac{2}{\pi}},\ \ \mathrm{E}(z'|P) = z. \tag{4.69}$$

   The population is pushed from the axis $0z$ to the straight line $w = \sigma\sqrt{\frac{2}{\pi}}$.

2. If the population P lies on the axis $0w$ (on the axis of symmetry $z = 0$), i.e. $x_1 = -x_2,\ q_1 = q_2$ then

$$\mathrm{E}(w'|P) = \sigma\sqrt{\frac{2}{\pi}} + \sigma\left(\phi\left(\frac{w}{\sigma}\right) + \frac{w}{\sigma}\,\Xi\left(\frac{w}{\sigma}\right)\right),\ \ \mathrm{E}(w'|P) = 0 \tag{4.70}$$

   and for large w we have

$$\mathrm{E}(w'|P) = \frac{1}{2}\left(\sigma\sqrt{\frac{2}{\pi}} + w\right). \tag{4.71}$$

3. If the population P lies on the straight line $w = z$, then x_2 is optimal i.e. $q_2 = f_{\max}$, and

$$E(w'|P) = \sigma\sqrt{\frac{2}{\pi}} + (1-\gamma^2)\,\sigma\left(\phi\left(\frac{w}{\sigma}\right) + \frac{w}{\sigma}\,\Xi\left(\frac{w}{\sigma}\right)\right), \qquad E(z'|P) = (1+\gamma)z. \tag{4.72}$$

   Because $\gamma < 0$ in this case, then the expected value of z' decreases. If q_2 is much larger than q_2 or $q_1 \to 0$, then $\gamma \to -1$ and

$$E(w'|P) = \sigma\sqrt{\frac{2}{\pi}}, \quad E(z'|P) = 0. \tag{4.73}$$

   The population P "jumps" to the location $\left(\sqrt{\frac{2}{\pi}}, 0\right)$ on the axis $0w$. The symmetric behavior may be observed for the population P from the line $w = -z$ ($q_1 = f_{\max}$).

4. For populations located far from both axes $0w$ and $0z$ if q_2 is much larger than q_2 we have

$$E(w'|P) = \sigma\sqrt{\frac{2}{\pi}}, \quad E(z'|P) = z - w. \tag{4.74}$$

All behavior described above is illustrated in Figure 4.9.

As can be seen, the straight line $w = \sigma\sqrt{\frac{2}{\pi}}$ plays a role of the attractor, sometimes called the *evolutionary channel*. If individuals in the population P are distant from each other (w is large) or $|q_1 - q_2|$ takes large values, then the expected behavior is the population "jump" to the evolutionary channel where the population individuals differ in $\sigma\sqrt{\frac{2}{\pi}}$. Next, the population slowly moves along the evolutionary channel with the step size $\Delta z = \pm\sigma\sqrt{\frac{2}{\pi}}$ until it bounds sufficiently close to the symmetry axis $z = 0$. The behavior described above exhibits the influence of the fitness function form. Because $|\gamma| \leq 1$ the coordinate z of the population state changes very slowly.

The shape of the regions in which the fitness strongly affects the expected value of the population state in the next genetic epoch also strongly depend on the fitness form. In such regions

$$E(w'|P) > \sigma\sqrt{\frac{2}{\pi}}, \quad E(\Delta z|P) < \sigma\sqrt{\frac{2}{\pi}}. \tag{4.75}$$

The formal analysis and discussion of the two-individual population's expected behavior is confirmed by tests performed for two fitness functions

$$f(x) = \exp(-5x^2) \tag{4.76}$$

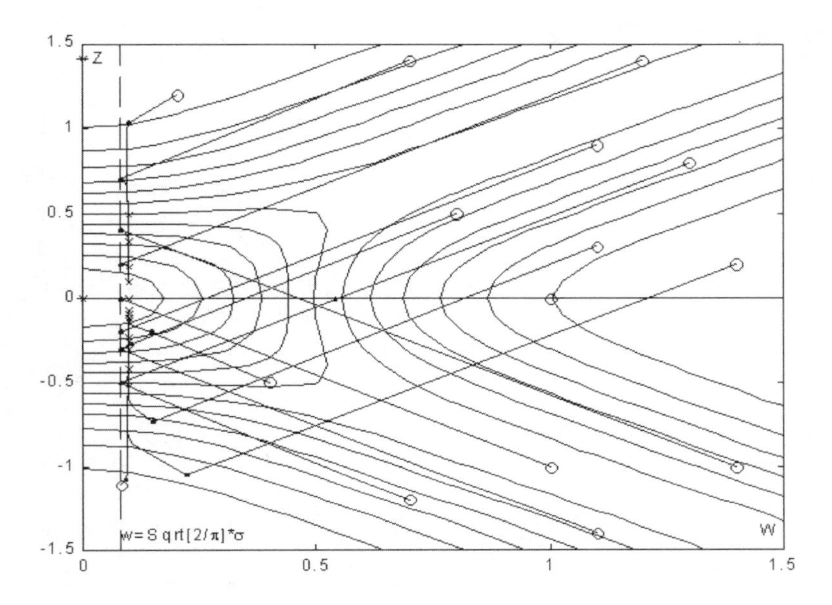

Fig. 4.10. Trajectories for the expected two-individual population in case of the unimodal fitness $f(x) = \exp(-5x^2)$. White circles mark the starting positions of populations, black dots the positions after the first epoch while crosses mark the positions after 20 epochs.

$$f(x) = \exp(-5x^2) - 2\exp(-5(x-1)^2) \tag{4.77}$$

In both cases we set $\sigma = 0.1$. The computation results for function 4.76 are presented in Figure 4.10 while for function 4.77 in Figure 4.11. We may observe that the evolution went according to the scheme of the expected position transformations described before. We may distinguish two evolution phases:

- The "jump" close to the identity axis $w = 0$ which is practically independent of the initial position. It is the effect of reproduction of only a single individual.
- Slow "drift" towards the maximizer of the fitness f. In the case of the unimodal function 4.76 this drift is immediate. If the fitness f is bimodal (see e.g. formula 4.77) such "drift" is much slower and may pass through the local extrema of f.

Another interesting subject is the study of the fixed points of the expected population operator, which correspond in some way to the fixed points of the genetic operator for the Simple Genetic Algorithm.

Let us turn back to more general fitness definitions 4.68. Such points may be obtained in this case by solving the system

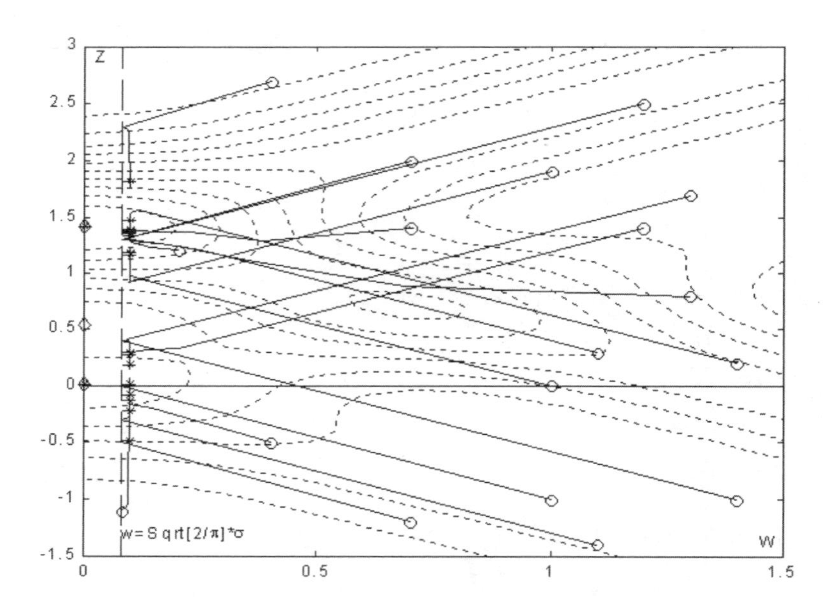

Fig. 4.11. Trajectories for the expected two-individual population in case of the bimodal fitness $f(x) = \exp(-5x^2) - 2\exp(-5(x-1)^2)$. White circles mark the starting positions of populations, black dots the positions after the first epoch while crosses mark the positions after 20 epochs.

$$\begin{aligned} w_s &= \mathrm{E}\left(w|P = (w_s, z_s)\right) \\ z_s &= \mathrm{E}\left(z|P = (w_s, z_s)\right) \end{aligned} \tag{4.78}$$

The second equation is satisfied if $\gamma = 0$, i.e. if $q_1 = q_2$ which gives

$$f\left(\frac{w_s + z}{\sqrt{2}}\right) = f\left(\frac{z - w_s}{\sqrt{2}}\right) \tag{4.79}$$

The approximated solution of the above equation equals $w_s = 0.97\sigma$ and does not depend on the fitness function, but only on the standard deviation of the mutation operation. However, the second coordinate z_s depends on fitness. If the fitness f is symmetric, then $z_s = 0$ which may be obtained from the equation 4.79. If fitness is symmetric and unimodal, then the point $(0.97\sigma, 0)$ is the only equilibrium point of the expected population (fixed point). In this equilibrium state both individuals are located at the same distance $\frac{w_s}{\sqrt{2}}$ from the global maximizer to f, so $(x_1)_s = -0.7\sigma$, $(x_2)_s = 0.7\sigma$.

For the function $f(x) = \exp(-ax^2)$ the point $(0.97\sigma, 0)$ is asymptotically stable if $a \in (0, a_0)$ where $a_0 = \frac{2}{(w_s)^2} = \frac{2}{(0.97\sigma)^2}$.

For multimodal functions we may obtain more fixed points of the expected population operator. In particular, for the function $f(x) = \exp(-a(x+d)^2) + \exp(-a(x-d)^2)$ we may have single or three fixed points. If $\sigma > 1.46d$ then the saddle point $(0.97\sigma, 0)$ is the only equilibrium of the expected two-individual population. If $\sigma > 1.46d$ then two more fixed points appear close to the local maximizer of the fitness, symmetrically to the axis $0w$. When σ decreases (the mutation becomes less intensive) then such points bound to the local maximizers. Generally, the process of two-individual population evolution may lead to finding symmetric maximizers if the standard deviation of the mutation σ is small in comparison to the distance between two maximizers.

4.3 The increment of the schemata cardinality in the single evolution epoch

The schemata theory was the first formal approach which tried to explain the asymptotic behavior of genetic algorithms with binary encoding. It was introduced by Holland in 1975 [85] and was devoted rather to the modeling of artificial life than to the stochastic global optimization algorithm analysis.

This approach has been criticized many times (see e.g. Grefenstette and Bayer [79], Grefenstette [78], Podsiadło [130]).

A significant improvement in the understanding of the main schemata theorem idea was delivered by Whitley in his Genetic Algorithm Tutorial [196]. A detailed explanation and exact formulation was given by Wright [204]. Recently, comments have been published by Kieś and Michalewicz [93] as well as by Reeves and Rowe [134] and Schaefer [151].

We intend to deliver quite a detailed and precise formulation and proof of these historical results in order to explain their true, very constrained meaning in genetic algorithm analysis. We will partially follow the way taken by the last group of cited authors and Vose's definition of the Simple Genetic Algorithm (see Vose [193], also Section 3.6).

Let us recall the binary genetic universum $\Omega = \{(a_0, a_1, \dots, a_{l-1});\ a_i \in \{0,1\},\ i = 0, 1, \dots, l-1\}$ which contains binary codes of the length l (see formula 3.5). The set $\mathcal{S} = \{0, 1, *\}^l$ will be called *the space of schemata.*

Definition 4.83. *The sequence $(h_0, h_1, \dots, h_{l-1}) \in \mathcal{S}$ depicts the schemata H which is the subset of the binary genetic universum ($H \subset \Omega$) given by the formula*

$$H = \left\{ (a_0, a_1, \dots, a_{l-1});\ a_i = \begin{cases} h_i & for\ h_i \in \{0,1\} \\ 0\, or 1 & for\ h_i = * \end{cases},\ i = 0, 1, \dots, l-1 \right\}.$$

Moreover, we assume that each schemata is nontrivial, i.e.

$$\exists i \in 0, 1, \dots, l-1;\ h_i \in 0, 1.$$

□

Each schemata may be characterized by some important parameters. The first of them is called *the schemata length* and the second *the degree of schemata.*

Definition 4.84. *The length $\Delta(H)$ of the schemata $H \subset \Omega$ is the maximum distance between the well-defined digits*

$$\Delta(H) = \max_{i,j=0,1,\dots,l-1} \{|i-j|;\ h_i, h_j \in \{0,1\}\}$$

while the degree $\aleph(H)$ of the schemata $H \subset \Omega$ is equal to the number of well-defined digits in H. □

It is easy to observe that $\Delta(H) \in \{0,1,\dots,l-1\}$ while $\aleph(H) \in \{1,\dots,l\}$.

We will model the finite population of individuals which are clones of elements from the genetic universum Ω as pairs $P = (\Omega, \eta)$, where the function $\eta : \Omega \to \mathbb{R}_+$ turns back the number of clones of the particular genotype from Ω (see definition 2.8).

The fitness function will be represented by the 2^l-dimensional vector $f = \{f_i\}, i \in \Omega$. We may denote for convenience by $\bar{f}_P$ *the mean fitness of the population* $P = (\Omega, \eta)$ i.e.

$$\bar{f}_P = \frac{1}{\mu} \sum_{i \in \Omega} f_i\, \eta(i) \tag{4.80}$$

where μ stands for the population cardinality $\mu = \#P < +\infty$.

Definition 4.85. *Let us consider the schemata $H \subset \Omega$. The schemata representation in the population $P = (\Omega, \eta)$ will be the multiset $rep(H,P) = (\Omega, \eta_{rep(H,P)})$ where*

$$\eta_{rep(H,P)}(i) = \begin{cases} \eta(i) & \text{if } i \in H \\ 0 & \text{otherwise} \end{cases}.$$

□

We now introduce the next two parameters of the schemata H related to the particular population P. The first of them will be *the mean fitness of the schemata representation*

$$\bar{f}_{rep(H,P)} = \frac{1}{\#rep(H,P)} \sum_{i \in \Omega} f_i\, \eta_{rep(H,P)}(i) \tag{4.81}$$

and *the fitness ratio of the schemata representation*

$$rat(H,P) = \frac{\bar{f}_{rep(H,P)}}{\bar{f}_P}. \tag{4.82}$$

Let us now consider the infinite set of populations $P_0, P_1, \ldots$ generated by the Simple Genetic Algorithm, so that $P_t = (\Omega, \eta_t)$ and $\#P_t = \mu < +\infty$ for all $t = 0, 1, \ldots$. We assume the multi-point mutation described by the formula 3.36 where the mutation mask is given by the probability distribution 3.37. We will restrict ourselves to the one-point crossover determined by formulas 3.40, 3.41 while the crossover type is given by formula 3.42.

The proportional selection distribution utilized in the SGA, described by formula 3.30 may be rewritten to the form

$$sel_f(\{i\}) = \frac{f_i\, \eta_t(i)}{\mu\, \bar{f}_P}. \tag{4.83}$$

Let us start with three lemmas that correspond to the Lemmas 3.1, 3.2 and 3.3 in Reeves and Rowe [134].

Lemma 4.86. *The probability of selecting the single individual that represents the schemata H at the epoch t equals*

$$\frac{1}{\mu}\, rat(H, P_t)\ \#rep(H, P_t).$$

□

Proof. The probability under consideration may be computed as follows

$$\sum_{i \in H} sel_f(\{i\}) = \frac{\sum_{i \in H} f_i\, \eta_t(i)}{\mu\, \bar{f}_P} = \frac{\sum_{i \in \Omega} f_i\, \eta_{rep(H,P)}(i)}{\mu\, \bar{f}_P} =$$

$$\frac{\#rep(H,P)}{\mu}\ \frac{\bar{f}_{rep(H,P)}}{\bar{f}_P} = \frac{\#rep(H,P)\ rat(H,P)}{\mu}.$$

It is obvious that the probability of selecting the individual that does not represent H is

$$1 - \frac{1}{\mu}\, rat(H, P_t)\ \#rep(H, P_t). \tag{4.84}$$

□

Lemma 4.87. *Let $x \in H$ and $y \in \Omega \setminus H$ (or $x \in \Omega \setminus H$ and $y \in \Omega$) be parent genotypes. The probability that the offspring $z \in \Omega$ obtained by the crossover of x and y belongs to H is larger than or equal to*

$$\frac{1}{2}\left(1 - p_c \frac{\Delta(H)}{l-1}\right).$$

□

Proof. The probability that x will be crossed non-trivially with y equals p_c because it is the probability that the crossover mask differs from the string composed of zeros or from the string composed of ones (see formula 3.41). Because the selection of the cutting position is uniform (see formula 3.42), then the conditional probability of destroying the schemata H in one child is less than or equal to $p_c \frac{\Delta(H)}{l-1}$. It may be substantially less because the part of the string y that is exchanged may also fit schemata, so this child may also belong to H. Anyway, the lower bound of the probability that this child belongs to H is $1 - p_c \frac{\Delta(H)}{l-1}$. From the Bayes rule applying for crossover we may obtain the following lower bound of the probability that the crossover result belongs to the schemata H

$$\Pr\{z \in H\} \geq \frac{1}{2}\left(1 - p_c \frac{\Delta(H)}{l-1}\right) + \frac{1}{2}\delta$$

where δ stands for the probability that the second child belongs to H. The thesis of Lemma 4.87 can be obtained by dropping the second term in the right-hand-side of the above inequality. □

Lemma 4.88. *Assuming $z \in H$ the probability that z', obtained by uniform mutation, stays in the schemata H is $(1 - p_m)^{\aleph(H)}$.* □

Proof. The digits predefined in the schemata H that occur in the string $z \in H$ will be passed to z' if the mutation mask has zero on the loci predefined in the schemata H. The probability of selecting such a mask due to the rule described in Section 3.5.1 equals the probability of sampling zeros on the $\aleph(H)$ positions and ones on the $l - \aleph(H)$ positions independently. Then we have $\Pr\{z' \in H\} = (1 - p_m)^{\aleph(H)} \, 1^{l-\aleph(H)}$. □

The next lemma will evaluate the probability of surviving the schemata in one-time sampling in the single step of SGA.

Lemma 4.89. *Assuming the current population P_t, the probability of one-time sampling the individual $z' \in H$ according to the SGA procreation rule is greater than or equal to*

$$\frac{rat(H, P_t) \; \#rep(H, P_t)}{\mu}$$

$$\left(1 - p_c \frac{\Delta(H)}{l-1}\left(1 - \frac{rat(H, P_t) \; \#rep(H, P_t)}{\mu}\right)\right)(1 - p_m)^{\aleph(H)}.$$

□

Proof. By using the Bayes rule we can evaluate

$$\Pr\{z' \in H\} = \Pr\{z' \in H | z \in H\} \Pr\{z \in H\} +$$
$$+ \Pr\{z' \in H | z \in \Omega \setminus H\} \Pr\{z \in \Omega \setminus H\}$$

where z' is obtained by mutation from the individual z.

Dropping the second term, which give us the probability of obtaining by mutation the string that belongs to H from the string that does not belong to H, and using the thesis of Lemma 4.88 we have

$$\Pr\{z' \in H\} \geq (1 - p_m)^{\aleph(H)} \Pr\{z \in H\}. \tag{4.85}$$

Let us now assume that the individual z was obtained from two parental strings x and y by one-point crossover. Using the Bayes rule once more we have

$$\begin{aligned}\Pr\{z \in H\} = & \Pr\{z \in H | x, y \in H\} \Pr\{x, y \in H\} + \\ & + \Pr\{z \in H | x \in H \text{ and } y \in \Omega \setminus H\} \Pr\{x \in H \text{ and } y \in \Omega \setminus H\} + \\ & + \Pr\{z \in H | x \in \Omega \setminus H \text{ and } y \in H\} \Pr\{x \in \Omega \setminus H \text{ and } y \in H\} + \\ & + \Pr\{z \in H | x, y \in \Omega \setminus H\} \Pr\{x, y \in \Omega \setminus H\}.\end{aligned}$$

Dropping the last term that expresses the probability of obtaining the string that belongs to the schemata H from two parents that do not belong to H, and next using the theses of Lemmas 4.86 and 4.87 we obtain

$$\Pr\{z \in H\} \geq 1 \cdot \left(\frac{rat(H, P_t) \; \#rep(H, P_t)}{\mu} \right)^2 + \frac{1}{2} \left(1 - p_c \frac{\Delta(H)}{l - 1} \right)$$

$$\left(\frac{rat(H, P_t) \; \#rep(H, P_t)}{\mu} \right) \left(1 - \frac{rat(H, P_t) \; \#rep(H, P_t)}{\mu} \right) +$$

$$\frac{1}{2} \left(1 - p_c \frac{\Delta(H)}{l - 1} \right) \left(1 - \frac{rat(H, P_t) \; \#rep(H, P_t)}{\mu} \right) \left(\frac{rat(H, P_t) \; \#rep(H, P_t)}{\mu} \right)$$

and now substituting to the inequality 4.85 we can complete the proof

$$\Pr\{z' \in H\} \geq \frac{rat(H, P_t) \; \#rep(H, P_t)}{\mu}$$

$$\left(1 - p_c \frac{\Delta(H)}{l - 1} \left(1 - \frac{rat(H, P_t) \; \#rep(H, P_t)}{\mu} \right) \right) (1 - p_m)^{\aleph(H)}.$$

□

Because the process of obtaining the next step population P_{t+1} is random, then the schemata representation cardinality $\#rep(H, P_{t+1})$ may also be handled as a random variable. Taking the above definitions and notations into account, the main results of this section may be formulated.

Theorem 4.90. *If all the assumptions made in Lemmas 4.86, 4.87, 4.88 are satisfied then we have*

$$E(\#rep(H, P_{t+1})) \geq rat(H, P_t) \; \#rep(H, P_t)$$

$$\left(1 - p_c \frac{\Delta(H)}{l - 1} \left(1 - \frac{rat(H, P_t) \; \#rep(H, P_t)}{\mu} \right) \right) (1 - p_m)^{\aleph(H)}.$$

□

Proof. The total procreation step in the SGA may be treated as the μ-time independent sampling of the single individual z' according to the Bernoulli scheme, in which the result $z' \in H$ is interpreted as the success. The number of successes in such sampling is the number of individuals in $rep(H, P_{t+1})$ i.e. the number of individuals in P_{t+1} that belong to the schemata H in the next epoch. Using the well-known formula for expectation in the Bernoulli scheme we obtain $\mathrm{E}(\#rep(H, P_{t+1})) = \mu \Pr\{z'\}$. It is sufficient to use Lemma 4.89 to complete the proof. □

The formula which was derived (in thesis of Theorem 4.90) is really nothing new and is almost the same as that given by Whitley [196]. The advantage of the above considerations seems to be the rigorous formulation and proof of all the steps, which underlines all their imperfections and simplifications leading in particular to underestimation of the expectation of the $rep(H, P_{t+1})$ cardinality.

The thesis of Theorem 4.90 inherits, of course, all the imperfections of the schemata theorem mentioned by many authors (see e.g. Reeves and Rowe [134], Whitley [196], Vose [193]. In particular:

- It is inaccurate, which means that it is possible to get better evaluations for $\mathrm{E}(\#rep(H, P_{t+1}))$ which do not involve the simplifications and underestimations that are visible in the proofs of Lemmas 4.87 and 4.89.
- It does not allow us to study the asymptotic behavior of the SGA. We need to be sure that P_t is the population in the t-th epoch in order to evaluate the statistics of $\#rep(H, P_{t+1})$ in the next step, so the formula could not be iterated along $t \to +\infty$. In order to study the asymptotic behavior it is necessary to evaluate the states or sampling probability distribution transition between consecutive epochs as is done in the Markov theory of SGA introduced by Vose.

The usual form of the schemata theorem, introduced by Holland [85] and reformulated by Goldberg [74], Michalewicz [110], Reeves and Rowe [134], brings the the following formula

$$\mathrm{E}(\#rep(H, P_{t+1})) \geq rat(H, P_t)\ \#rep(H, P_t) \left(1 - p_c \frac{\Delta(H)}{l-1} - \aleph(H)\, p_m\right).$$

The above evaluation is weaker than the one presented in Theorem 4.90. Really, taking into account the inequality $(1 - p_m)^{\aleph(H)} \geq 1 - \aleph(H)\, p_m$ we obtain

$$1 - p_c \frac{\Delta(H)}{l-1} \left(1 - \frac{rat(H, P_t)\ \#rep(H, P_t)}{\mu}\right) =$$

$$1 - p_c \frac{\Delta(H)}{l-1} + p_c \frac{\Delta(H)}{l-1}\, \frac{rat(H, P_t)\ \#rep(H, P_t)}{\mu} \geq 1 - p_c \frac{\Delta(H)}{l-1}$$

so, then

$$rat(H,P_t)\ \#rep(H,P_t)\left(1-p_c\,\frac{\Delta(H)}{l-1}\left(1-\frac{rat(H,P_t)\ \#rep(H,P_t)}{\mu}\right)\right)$$

$$(1-p_m)^{\aleph(H)} \geq rat(H,P_t)\ \#rep(H,P_t)\left(1-p_c\,\frac{\Delta(H)}{l-1}\right)(1-\aleph(H)\,p_m) \geq$$

$$rat(H,P_t)\ \#rep(H,P_t)\left(1-p_c\,\frac{\Delta(H)}{l-1}-\aleph(H)\,p_m+p_c\,\frac{\Delta(H)}{l-1}\aleph(H)\,p_m\right) \geq$$

$$rat(H,P_t)\ \#rep(H,P_t)\left(1-p_c\,\frac{\Delta(H)}{l-1}-\aleph(H)\,p_m\right).$$

4.4 Summary of practicals coming from asymptotic theory

Significant asymptotic theory results have been obtained mainly for genetic algorithms whose behavior may be described by the trajectory in the space of states composed of populations (e.g. $\mathcal{E} = U^{\mu}/eqp$ for constant, finite size $\mu < +\infty$ population algorithms) or their unambiguous representations (e.g. $\mathcal{E} = \Lambda^{r-1}$ for SGA). In such cases, under some additional conditions (see formula 4.6), the stochastic dynamics of such algorithms may be modeled by Markow chains. The effective analysis of the asymptotic behavior by $t \rightarrow +\infty$ is easier if the probability transition rule $\tau : \mathcal{E} \rightarrow \mathcal{M}(\mathcal{E})$ is stationary i.e. does not depend on the epoch number. It holds if selection and genetic operations do not depend explicitly or implicitly on the epoch number t. This is the case in all the genetic algorithm instances effectively analyzed in this chapter.

The main feature which was studied is the ergodicity of the Markov processes that can model particular groups of genetic algorithms. Ergodicity roughly means the lack of absorbing states that may be occupied by the genetic algorithm infinitely and the guarantee of visiting all the states in $\mathcal{E}$.

Positive results for the SGA (see Theorem 4.31) and for the evolutionary algorithm $(\lambda + \mu)$-type with elitist selection (see formula 4.52) were obtained assuming the rather "brutal" condition that forces strictly positive mutation in every genetic epoch (the mutation rate is positive $p_m > 0$ in the case of the SGA, and the strictly positive mutation operation kernel $\tau_{mut}(x)(\{x'\}) > 0,\ \forall x, x' \in U$ in the case of the EA $(\lambda + \mu)$-type with elitist selection). In the case of the EA we have to additionally assume that the space of states is finite ($\#\mathcal{E} < +\infty$). Such an assumption guarantees a much stronger feature, the algorithm can pass between two arbitrary stages from $\mathcal{E}$ in a single genetic epoch with a strictly positive probability. Summing up this part of results we may say that:

If the genetic algorithm can be modeled by the ergodic Markov chain with states in $\mathcal{E} = U^{\mu}/eqp$ then it passes with all possible populations, in particular those that contain individuals whose phenotypes correspond to global or local *extrema of the objective function (or the best approximations in the set of phenotypes $\mathcal{D}_r$). Such behavior is precisely called the asymptotic correctness in the probabilistic sense and the asymptotic guarantee of success (see definitions 2.16, 2.17).*

Ergodicity of the Markov chain that models the genetic algorithm also forces two kinds of convergence for two important classes of algorithms:

The Simple Genetic Algorithm with finite population ($\mu < +\infty$) *case.* The weak convergence of the sequence of measures $\{\pi^t_\mu\}$, $\pi^t_\mu \in \mathcal{M}(\mathcal{E})$, $t = 0, 1, \ldots$ to the limit, invariant measure π_μ for $t \rightarrow +\infty$ was proved (see Theorem 4.31). Moreover, the limit measure π_μ is strictly positive and does not depend on the starting measure π^0_μ.

Let us recall that the measure π^t_μ determines the probability of selecting a new population from the space $\mathcal{E}$ in the t-th genetic epoch. The progress in the $\{\pi^t_\mu\}$ convergence illustrates the process of SGA learning. The algorithm gathers information about the optimization problem, which is memorized in the measure π^t_μ. The ergodicity guarantees the stable convergence of the learning process, but does not guarantee the total, maximum level of knowledge to be gathered.

Please note that the convergence of the sequence $\{\pi^t_\mu\}$ does not necessarily imply the stabilization of the population state by $t \rightarrow +\infty$. This feature is surprising even for many researchers that apply genetic algorithms. It may be stressed in the form of the so-called *genetic paradox*:

If the Simple Genetic Algorithm is convergent (the sequence $\{\pi^t_\mu\}$ converges) then it it not "convergent" i.e. all individuals do not lead to the global maximizer. However, if the SGA populations converge deterministically to the monochromatic one then the related Markov chain has the absorbing state and then is not ergodic. The global search ability in $\mathcal{D}_r$ is lost (the algorithm does not posses the asymptotic correctness in the probabilistic sense and the asymptotic guarantee of success).

The case of evolutionary algorithm ($\mu + \lambda$) *with elitist selection.* The convergence of maximum fitness appearing in the consecutive population to the global fitness maximum under rather restrictive assumptions has been proved (see Theorem 4.70, example 4.72).

The next results of the practical meaning have been obtained for the Simple Genetic Algorithm instances for which the genetic operator G is focusing (see definition 4.34) and they are "well tuned" with respect to the group of local extremes of the objective functions to be searched (see definition 4.63). For such algorithms the sampling measure becomes dense in the central parts of the basins of attraction of local extremes (see Theorem 4.66 and Remark 4.67) if populations are sufficiently large and a sufficiently large number of genetic epochs are passed. In such cases the information about the evolutionary landscape may be drawn by the a'posteriori analysis of the counting measure obtained from the final population (see remark 4.59). The simplest possibility of such a process, which consists of finding the level sets of specially defined density of the above mentioned measure (Clustered Genetic Search), will be presented in Section 6.3.1. Well tuned genetic algorithms may also constitute an effective tool for producing the random sample in the first step of the two-phase global optimization stochastic strategies (see Section 6.3).

The *rate of convergence* of genetic algorithms is generally difficult to evaluate analytically. The logarithmic rate of convergence with respect to the number of genetic epochs of the idealized, infinite population SGA to the fixed points of the genetic operator G was delivered by Theorem 4.53. The evaluation of the SGA efficiency, restricted to the single evolution step that consists of passing between two consecutive genetic epochs, may be drawn from the schemata Theorem 4.90. In particular, the one-step increment of the population sub-multisets, which contain solutions to global optimization problem $\mathbf{\Pi_1}$ may be statistically estimated. The next important results in this area (see remark 4.79) prove the geometrical rate of convergence of the maximum fitness value evaluation with respect to the number of genetic epochs of the $(\mu + \lambda)$ evolutionary algorithm. Very strong assumptions with respect to the fitness increment between two consecutive genetic epochs were accepted. The mean hitting time of the global maximizer for the special kind of random walk (genetic algorithm with a single individual population, AB mutation and hard selection) assuming the arbitrary unimodal fitness can also be evaluated (see Theorem 4.81 and remark 4.82).

The next important feature of genetic algorithms for which mathematically-verified results are rather rare is the *stopping rule*. It is worth mentioning the results of Hulin [87] who tried to construct the stopping rule of the Bayes type, typical for the Monte Carlo algorithms. Another type of effective, mathematically-verified stop criterion for the two-phase global optimization strategy that utilizes the well tuned Simple Genetic Algorithm and the fitness landscape erosion technique (hill crunching described in Section 5.3.5) will be delivered in Sections 6.3.2.

5

Adaptation in genetic search

The previous Chapters 3, 4 presented perhaps the simplest instances of genetic global optimization algorithms that only exploit basic mechanisms of genetic computation, such as mutation, crossover and selection. All these mechanisms do not change with respect to the genetic epoch and are "blind" to the optimization problem to be solved as well as to the knowledge about it currently gathered by the algorithm. Such simplicity allows us to construct the mathematical models and perform deep formal analysis of the asymptotic behavior, which is helpful in understanding the real nature of genetic global optimization. However, the efficiency of basic genetic search mechanisms are frequently criticized.

This chapter discusses the adaptation techniques and strategies that go toward the better efficiency of global genetic searches in continuous domains. The first group of them consists of on-line modifying genetic mechanisms during the genetic epoch progress, according to the assumed plan or according to the feedback coming from one or more previous steps (see Section 5.3). The second group introduces more sophisticated multi-deme searches that allow concurrent checking of the whole admissible domain (see Section 5.4).

5.1 Adaptation and self-adaptation in genetic search

The most basic reference algorithm that is worth mentioning in the taxonomy of adaptive stochastic global optimization searches is called *Pure Random Search* (PRS) (see Section 2.2). It involves the generation of all populations $\{P_t\}$, $t = 0, 1, \ldots$ according to the same, uniform probability distribution over the admissible domain $\mathcal{D}$.

Such an algorithm, in which the sampling probability distribution is still or varies only with respect to the deterministic, a'priori assumed plan, may be called *devoid of the adaptation mechanisms.*

The classical genetic algorithms described in Chapters 3, 4, perform the sampling distribution modification in each genetic epoch. All these algorithms

R. Schaefer: *Foundation of Global Genetic Optimization*, Studies in Computational Intelligence (SCI) **74**, 115–152 (2007)
www.springerlink.com

can be modeled by uniform Markov chains with the space of states $\mathcal{E}$ which represent all the possible populations and with the stationary transition rule $\tau : \mathcal{E} \rightarrow \mathcal{M}(\mathcal{E})$ which does not depend on the genetic epoch $t = 0, 1, \ldots$ as well as from earlier states. More precisely, the form of the mapping τ in the tth epoch depends neither on t nor on the states $x^0, x^1, \ldots, x^{t-1} \in \mathcal{E}$ (see Section 4.1.1). However, the probability distribution $\pi^{t+1} = \tau(x^{t+1})$ which is used for the next population sampling (see Figure 4.1) varies in time because $\tau(x^{t+1}) \neq \tau(x^t), t = 0, 1, 2, \ldots$. It has an obvious influence on the probability distributions which are used when sampling individuals (and then their phenotypes) from the genetic universum U (from the admissible domain $\mathcal{D}$). For the Simple Genetic Algorithm the consecutive samples from $\mathcal{D}$ were selected according to the probability distributions $\hat{\theta}'(x^t) \in \mathcal{M}(\mathcal{D})$ (see formula 4.26 and Remark 4.59). The form of the mapping $\hat{\theta}'$ does not depend on the epoch counter t or on the sampling results in previous epochs. A similar scheme was accepted in the case of classical evolutionary algorithms, as mentioned in Sections 3.7 and 4.1.3.

Summing up, classical genetic algorithms involve the adaptation mechanism that modifies the probability distribution, which is utilized by sampling the consecutive multiset from the admissible domain $\mathcal{D}$ (phenotypes assigned to the next step population). This mechanism depends only on the assumed genetic operations and the selection operation as well as their constant parameters (e.g. the mutation rate p_m and the crossover rate and type $p_c, type$ for the Simple Genetic Algorithm described in Sections 3.5, 3.6). This adaptation does not need any interference during the algorithm iteration, so we will call it *self-adaptive genetic algorithm.*

In contrast to the self-adaptive, classical genetic algorithms the *adaptive stochastic search* and the *adaptive genetic algorithm* perform single-step sampling from $\mathcal{D}$ by using mechanisms that are intentionally modified during the computation. In particular, the genetic operations, selection and their parameters settings may change in the consecutive genetic epochs. Moreover, the number of individuals that is processed in the particular epoch as well as other features of algorithms may be modified.

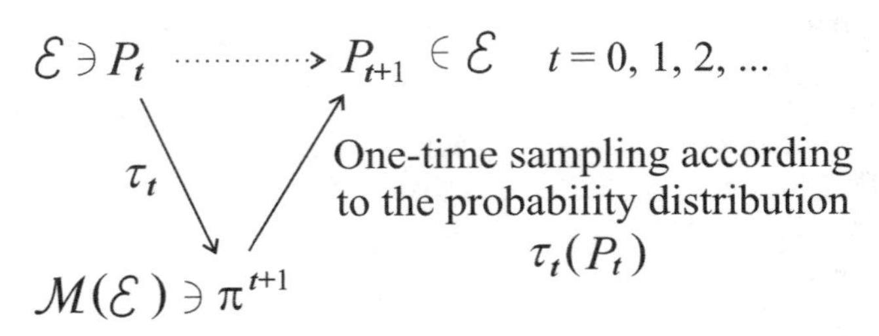

Fig. 5.1. Sampling scheme for the adaptive genetic algorithms.

In the case of adaptive stochastic searches, the possibility of the dynamic behavior modeling by using a uniform Markov chain (i.e. the Markov chain with a stationary probability transition rule τ) is lost. If the adaptive genetic strategy allows the precise definition of the space of states $\mathcal{E}$, then the single-step progress may be expressed by the diagram 5.1, similarly to the diagram 4.1 which describes the sampling rule in the case of classical genetic algorithms. The transition function τ_t that determines the probability distribution π^{t+1}, which is used by the next epoch population sampling may now depend on the current genetic epoch counter t, the current population P_t, earlier populations produced by the algorithm $P_0, P_1, \ldots, P_{t-1}$ as well as on some parameter vector $u(t)$ that controls the evolutionary process. Entries of the vector $u(t)$ may have an influence on the selection and genetic operations as well as on other parameters, such as the population cardinality. The relaxation of the sampling rule may result not only in the loss of its stationarity but also in the loss of Markovian principles in general (see formula 4.6).

5.2 The taxonomy of adaptive genetic strategies

The main goal of adaptation, which is introduced in the sampling rule of genetic search, is to increase the total efficiency of such strategies. Such efficiency improvement may be reached by elimination of basic disadvantageous behavior observed during the virtual evolution. Disadvantages under consideration may fall into three groups:

1. Slow concentration of individuals in the neighborhood of the global maximizer as well as slow bounding of the best fitted individual to this maximizer.
2. Small range of penetration of the individual phenotypes in the domain $\mathcal{D}$ which only reach the basins of attraction of several local maximizers after a large number of genetic epochs.
3. Evolutionary cliffs (or evolutionary traps) which result in the long-term occupation of the basin of attraction of a single local maximizer by almost all individuals.

There are numerous adaptive genetic strategies described in monographs (see e.g. Goldberg [74] Michalewicz [110], Michalewicz, Bäck, Fogel [15], [10], [11] and in a large number of research papers. It is difficult to introduce an exhaustive taxonomy of such solutions. Some very interesting, and perhaps the most comprehensive, approaches to the classification of adaptive genetic strategies may be found in Michalewicz, Bäck, Fogel monographs [15], [10], [11] and also in the book written by Arabas [5]. A taxonomy, restricted to strategies dedicated to solving continuous global optimization problems with a multimodal objective function, was presented by Kołodziej [96]. The reasons that make this venture so difficult are:

- the complex nature of many adaptive strategies,
- the similar final effect obtained by some strategies constructed in order to satisfy quite different principles.

Both of the above mean there are a great deal of solutions already published by researchers and make it difficult to establish a single, uniform criterion of adaptive strategy classification. We decide to introduce two criteria and then two partially-dependent taxonomies. The first and simplest of them takes *goals of adaptation* into account, while the second one differentiates adaptation strategies with respect to *elementary techniques to be utilized.* The presented taxonomies do not aspire to be general or exhaustive ones. According to restrictions accepted in this book, these taxonomies omit adaptation techniques that can only be applied in the case of solving discrete optimization problems. We do not include strategies that allow us to fit the virtual evolution to the dynamically changing fitness either.

Concerning the first established criterion, the main goals that may be distinguished in the genetic adaptive searches applied to the continuous global optimization problems are the following:

A. Strengthen and fasten local search ability. Activities that are performed result in changing the genetic algorithm behavior in the area of attraction set of the single, but arbitrary local maximizer of the objective function Φ. They generally lead to an increase in the sampling measure density in the central part of this attraction set, and as a consequence quicken the sampling of the individual with the phenotype close to this local maximizer. In the case of multi-deme searches it is possible to attain this goal concurrently in attraction sets of more than one local maximizer which corresponds to the goal C.

B. Increment of the population discrepancy and mobility. Particular targets which such techniques try to attain are directed or undirected changes in the sampling measure density leading to intensive, chaotic penetration of the whole admissible domain $\mathcal{D}$ or the migration of the population to the attraction set of another, neighboring local maximizer. The benefits of the periodical application of such strategies is the easier passing of evolutionary cliffs, i.a. capturing the attraction set of the local maximizer with the higher value of the objective function Φ.

C. Total concurrent search. These strategies force the concurrent increase of the sampling measure density in the central parts of attracting sets of many local maximizers of Φ. These measures are utilized to sample the single population or many demes by genetic algorithms in the consecutive steps of evolution.

The second taxonomy has a tree form (see Figures 5.2, 5.3, 5.4). Its root corresponds to the classical genetic algorithms that utilize only selection, mutation and crossover operations which are fixed in time. The population

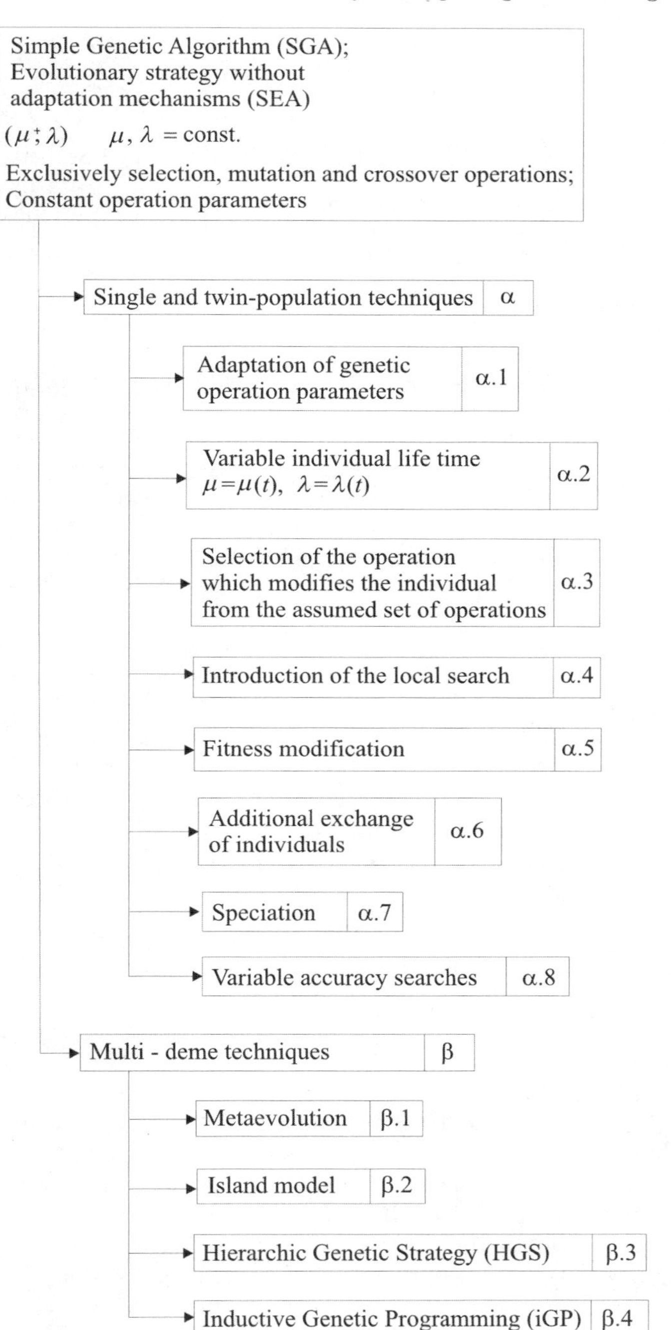

Fig. 5.2. The tree of adaptation techniques. Part 1.

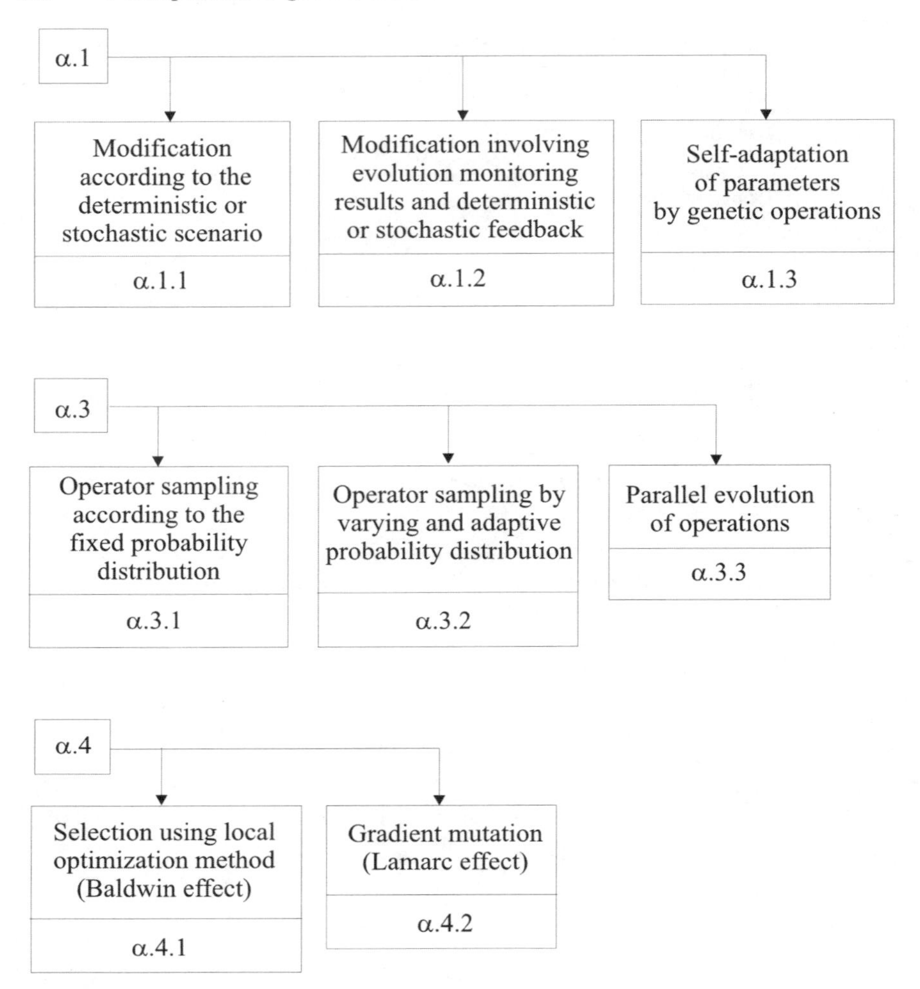

Fig. 5.3. The tree of adaptation techniques. Part 2.

cardinality as well as the succession scheme are also fixed ($\mu, \lambda = const.$, see 3.9). Leafs of the classification tree constitute classes of elementary adaptation techniques. In engineering practise, we observe complex strategies that involve elementary techniques from several groups in order to reach one of the goals A., B. and C. or the weighted combination of more than a single goal.

The basic criterion that allows us to spread the root into two boughs is the structure of the random sample. It contains single- and twin-population techniques and multi-deme techniques with the extended relation structure between respective populations. The following sections contain a short description of more than twenty selected elementary adaptation techniques and important examples of their application.

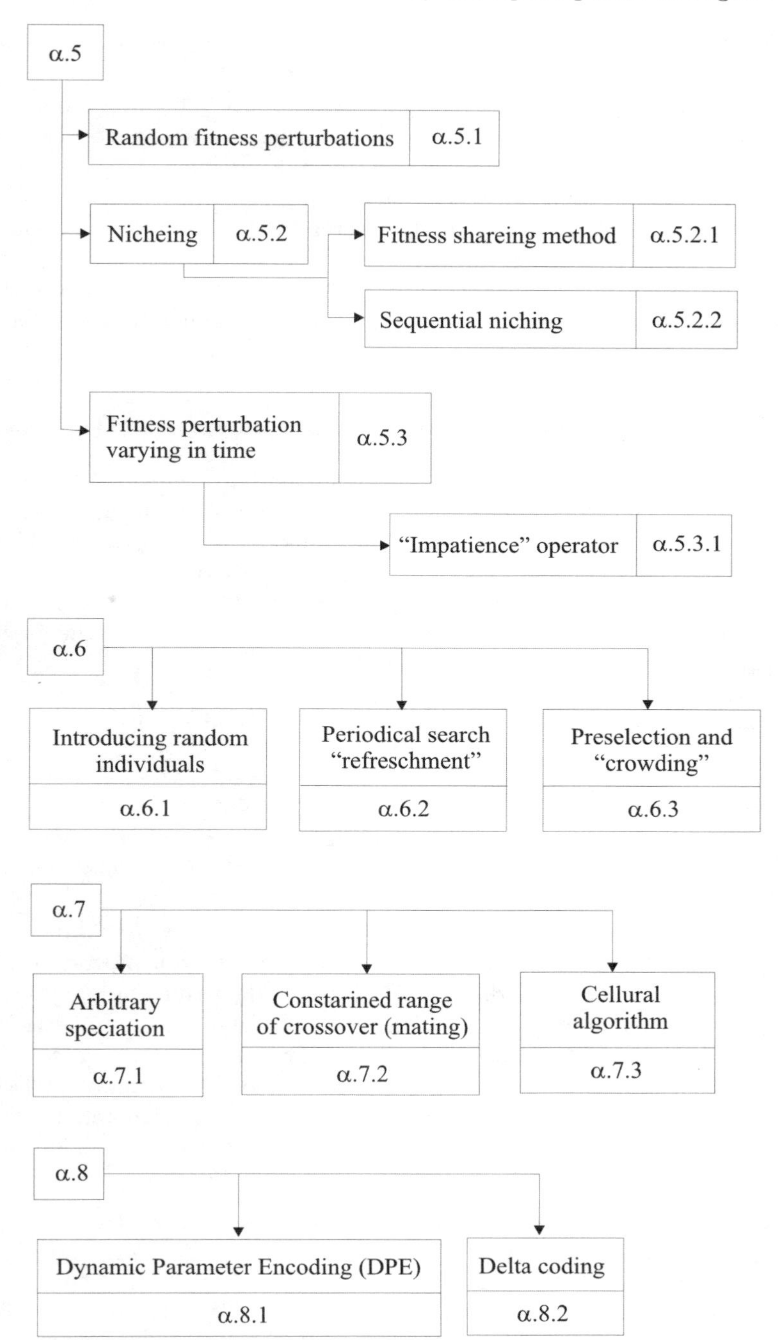

Fig. 5.4. The tree of adaptation techniques. Part 3.

5.3 Single- and twin-population strategies (α)

Single population adaptive strategies are historically the earliest ones. They intend to modify the population dynamics in order to reach one or more of the goals A., B., or C. together. In the case of multiple goals, single-population strategies try to reach them consecutively.

Twin-population strategies usually differentiate the role of each population evolving concurrently. One of them is the typical population of encoding candidates of solutions, e.g. their phenotypes belong to the admissible domain $\mathcal{D}$. The second population plays a control rule e.g. dynamically profiles genetic operations or/and the selection process.

5.3.1 Adaptation of genetic operation parameters (α.1)

Adaptation of genetic operation parameters is perhaps the simplest and most frequently utilized class of adaptation genetic strategies. Its formal description is based on the extending of the individual model to the pair (x, s), where $x \in U$ stands for the individual's genotype and $s \in Q \subset \mathbb{R}^k, k \in \mathbb{N}$ is the vector of genetic operation parameters (see Bäck [13]). The set of all possible parameter values Q is usually bounded and regular (e.g. has the Lipschitz boundary).

Let us denote by $\{g_i\}$ genetic operations that can affect population genotypes. Each of them may be formalized as the random function

$$g_i : U^p \times Q \to \mathcal{M}(U), \quad p \in \mathbb{N} \tag{5.1}$$

where p equals 1 or 2 for typical operations (mutation and crossover). Only the operations taken at the parameter value s, i.e. $\{g_i(\cdot, s)\}$ may act on the individual represented by the pair (x, s).

If the algorithm does not allow adaptation, then $s = const.$ which means it is this same for all genotypes and does not change during the evolution. For example, in the case of the Simple Genetic Algorithm the parameter set $Q \subset \mathbb{R}^{r+2}$ and equals $Q = [0, 1] \times [0, 1] \times \Lambda^{r-1}$ while the parameter vector $Q \ni s = (p_m, p_c, type)$ is composed of mutation and crossover rates and the crossover type vector (see Sections 3.5.1, 3.5.2 and 4.1 for details). Because the SGA does not allow any adaptation, each of these parameters is valid for all genotypes and does not change with respect to the genetic epoch counter.

Parameters modification according to the deterministic and stochastic rules (α.1.1)

In these techniques, the parameter vector s usually has the same value for all individuals in the single epoch population, but may be modified when passing between two consecutive genetic epochs.

The parameter which is most frequently adopted is the parameter that controls the mutation intensity. It may be, in particular, the rate p_m of the single bit mutation in the case of genetic algorithms using binary encoding (see e.g. formulas 3.36, 3.37 for SGA) and the standard deviation σ or the covariance matrix $\mathbf{C}$ for the evolutionary algorithms with normal mutation (see formula 3.51). Deterministic formulas that define such modification were given by Fogel [68], Lis [103], [104], Michalewicz [110], Bäck [12] and Arabas [5].

The simplest, widely-cited formula of this type, that may adapt the mutation rate by passing to the next genetic epoch in the case of binary mutation, has the form

$$p'_m = \alpha p_m, \ \alpha \in [0, 1] \tag{5.2}$$

where α stands for its constant parameter. Bäck in [14] delivers another formula which determines the mutation rate at the tth genetic epoch.

$$p_m(t) = \frac{1}{240} + \frac{0.11375}{2^t} \tag{5.3}$$

Such formulas decrease the mutation intensity in consecutive genetic epochs. This results in increasing the selection pressure and in strengthening the local search ability in the later genetic epochs which follow a more chaotic search in the introductory period of evolution.

Bäck and Schütz [16] suggested the stochastic rule of the binary mutation rate adaptation.

$$p'_m = \left(1 + \frac{1 - p_m}{p_m} \exp(-0.2\mathcal{N}(0,1))\right)^{-1} \tag{5.4}$$

where $\mathcal{N}(e, \sigma)$ denotes the real value random variable with the normal distribution, mean e and standard deviation σ. This formula is applicable in the case of maximizing simple objective functions.

Let us consider the $(\mu + \lambda)$ evolutionary algorithm with phenotypic encoding and normal mutation. If the admissible domain $\mathcal{D} \subset \mathbb{R}^N$, then we may denote $x = \{x_i\}, s = \{\sigma_i\}, \ i = 1, \dots, N$ and $\mathbf{C} = diag\{\sigma_i\}$ is the covariance matrix used by the mutation operation. Schwefel [163], Schwefel and Rudolph [165], Bäck [14] and Arabas [5] reported the rule that adopts the "range of mutation"

$$\begin{cases} x'_i = x_i + \sigma'_i(\mathcal{N}(0,1))_i, \ i = 1, \dots, N \\ \sigma'_i = \sigma_i(\alpha\mathcal{N}(0,1) + \beta(\mathcal{N}(0,1))_i) \\ \alpha = \dfrac{K}{\sqrt{2N}}, \ \beta = \dfrac{K}{\sqrt{2\sqrt{N}}} \end{cases} \tag{5.5}$$

where the parameter K is called normalized rate of convergence. Its value can be set 1 at the start of evolution. A much more extended formula including the correlations among the individual coordinates (the wholly filled covariance matrix $\mathbf{C} = \{c_{ij}\}$) was given by Schwefel [163].

$$\begin{cases} x'_i = x_i + \mathcal{N}(0, \mathbf{C}),\ i = 1, \dots, N \\ c_{i,j} = \frac{1}{2} \tan(2a'_k)((\sigma'_i)^2 - (\sigma'_j)^2),\ i \neq j, \\ k = \frac{1}{2}(2N - i)(i + 1) - 2N - j \\ c_{ii} = \sigma'_i \\ \sigma'_i = \sigma_i(\alpha \mathcal{N}(0, 1) + \beta(\mathcal{N}(0, 1))_i) \\ a'_i = a_i + 0.0873(\mathcal{N}(0, 1))_i \\ \alpha = \frac{K}{\sqrt{2N}},\quad \beta = \frac{K}{\sqrt{2\sqrt{N}}} \end{cases} \tag{5.6}$$

In the above formulas, the random variables $(\mathcal{N}(0, 1))_i$ are independent instances of the normal random variable $\mathcal{N}(0, 1)$ (independently sampled) while $\mathcal{N}(0, \mathbf{C})$ stands for the N-dimensional random variable with the normal probability distribution, 0 mean and the covariance matrix $\mathbf{C}$. Its density is given by the formula

$$\rho_{\mathcal{N}(0,\mathbf{C})}(x) = \frac{\exp\left(-\frac{1}{2} x^T \mathbf{C} x\right)}{((2\pi)^N \det(\mathbf{C}))^{\frac{1}{2}}} \tag{5.7}$$

The parameters σ_i, a_i have to be initialized at the start of evolution. The coefficients α, β called "learning rates" may also be computed by using other formulas recommended by Schwefel [163] and Beyer [25]. In this case, the parameter vector assigned to each individual with the genotype x equals $s = (\{\sigma_i\}, \{a_k\}),\ i = 1, \dots, N,\ k = 1, \dots, \frac{N(N-1)}{2}$. Please note that the parameter vector value s may vary for different individuals in the population.

All techniques adapting the mutation parameters described above in this section try to satisfy the goal A.

Methods that try to improve the evolution by crossover parameter modification generally go in two directions:

- They prevent individuals whose phenotypes are located too closely to each other in the admissible domain $\mathcal{D}$ from crossing. The phenotype distance is measured using the distance function d in the solution space V and should be not less than the assumed real, positive constant c. This constant can be modified proportionally to the inverse of the genetic epoch counter t (e.a. proportionally to $\frac{1}{t}$).
- They prevent incest. Several levels of the genealogical tree of each individual are memorized. Crossover is prevented if parents have a common

ancestor in the last k levels of their trees. The integer k stands for the parameter of such a strategy.

The methods of adaptation of the rate of crossover p_c in the binary crossover and, more generally, the modification of probability distribution of the crossover mask were given by Schaffer and Morishima [159].

For the $(\mu + \lambda)$ strategy, crossover parameter modification was studied by Rzewuski, Szreter and Arabas [146]. Now we assume the same parameter vector $s = \{\sigma_i\},\ i = 1, \ldots, N$ for the individual $x = \{x_i\}$, as used for mutation (see formulas 5.5, 5.6).

They suggest modifying the arithmetic crossover of two parental individuals $x^1, x^2 \in \mathcal{D} \subset \mathbb{R}^N$ to the form:

$$\begin{cases} x'_i = x^1_i + \gamma \mathcal{N}(0,1)(x^2_i - x^1_i),\ i = 1, \ldots, N \\ \gamma = \dfrac{\rho_{\mathcal{N}(x^1, \mathbf{C}^2)}(x^2)}{\rho_{\mathcal{N}(x^2, \mathbf{C}^1)}(x^1)} \\ \mathbf{C}^j = \{(\sigma'_i)^j\, \delta_{ik}\},\ j = 1, 2 \end{cases} \tag{5.8}$$

where the vector $x' = \{x'_i\}$ is the child of x^1, x^2. The density functions $\rho_{\mathcal{N}(x^i, \mathbf{C}^j)},\ i, j = 1, 2$ are given by the formula 5.7. The parameters $s'^j = \{(\sigma'_i)^j\},\ j = 1, 2,\ i = 1, \ldots, N$ are modified in the same way as in the formula 5.5.

Another formula for adaptive crossover was delivered by Arabas [5]:

$$\begin{cases} x' = \alpha x^1 + (1 - \alpha) x^2,\ x'' = \alpha x^2 + (1 - \alpha) x^1 \\ (\sigma')^1 = \alpha \sigma^1 + (1 - \alpha) \sigma^2,\ (\sigma')^2 = \alpha \sigma^2 + (1 - \alpha) \sigma^1 \\ \alpha = \mathcal{N}(0, 1) \end{cases} \tag{5.9}$$

where $\sigma^j = \{(\sigma'_i)^j\},\ j = 1, 2$. Both adaptive crossover rules described above belong to class B. because they lead to the discrepancy increasing in consecutive genetic epochs.

Parameter modification using monitoring of the evolutionary process (α.1.2)

In order to modify genetic operation parameters, while respecting the monitoring results of the evolutionary process, the proper state parameters and state dynamic parameters have to be distinguished and some characteristic behavior has to be indicated and classified.

Let us consider the evolutionary strategy with phenotypic encoding and the admissible domain of search $\mathcal{D} \subset \mathbb{R}^N$ as mentioned in the previous section. For its simplest instance $(1 + 1)$ (random walk) we may use the adaptation strategy called $\frac{1}{5}$ – rule, introduced by Rechenberg (see Arabas [5]). Now we consider only the single parameter $s = \sigma$ which is the standard deviation that controls the mutation operation.

$$\begin{cases} x'_i = x_i + (\mathcal{N}(0,\sigma'))_i, \; i = 1,\ldots,N \\ \sigma' = \begin{cases} c_I\,\sigma & \text{if } f(x') < f(x) \text{ holds at least} \\ & \dfrac{k}{5} - \text{times in consecutive k epochs} \\ c_D\,\sigma & \text{in other case} \end{cases} \end{cases} \tag{5.10}$$

Real numbers $c_I > 1$, $c_D < 1$ and the positive integer k stand for the parameters of this strategy.

The next two examples of evolutionary strategies which introduce feedback based on evolution monitoring are called *Evolutionary Search with Soft Selection* (ESSS). Both of them are based on the classical evolutionary $(\mu+\lambda)$ mechanism and phenotypic encoding, as in the previous case. The adaptation mechanism is activated if the evolving population satisfies the "trap test" at some evolutionary step k. The two following trap tests were defined by Obuchowicz and Patan [118], [122]:

1. Mean fitness in the population growth is less than $p\%$ in the last k_{ntrap} epochs.
2. The displacement norm of the expected value of population phenotypes (the norm of the population "mass" center displacement) is less than σ in the last k_{ntrap} epochs.

The first strategy called *Evolutionary Search with Soft Selection - Simple Variation Adaption* (ESSS-SVA), increases the standard deviation of the mutation operation in each epoch that follows the trap test fulfillment, i.e.

$$\sigma' = \alpha\,\sigma, \quad \alpha > 1 \tag{5.11}$$

If the trap test is not satisfied, the standard deviation σ is set to its initial value. The constant α stands for the ESSS-SVA parameter.

The second strategy of this group, called *Evolutionary Search with Soft Selection - Forced Direction of Mutation* (ESSS-FDM), changes the mutation operation after the trap test is satisfied at the genetic epoch t. The temporary mutation operation has the form:

$$\begin{cases} x'_i = x_i + (\mathcal{N}(m_i,\sigma)), \; i = 1,\ldots,N \\ m_i = \zeta\,\sigma\,\dfrac{E_i(P_t) - E_i(P_{t-1})}{\|E_i(P_t) - E_i(P_{t-1})\|} \\ E_i(P_t) = \dfrac{1}{\#P_t}\sum_{x\in P_t} \eta_t(x)\,x_i \end{cases} \tag{5.12}$$

where $P_t = (\mathcal{D}, \eta_t)$ is the population representation. The real number ζ stands for the strategy parameter.

Both ESSS strategies described above mobilize the population if it occupies the basin of attraction of the single local extreme for a long time (the trap test is positively verified). ESSS-SVA temporary increases the chaotic search component which enlarges the region effectively checked while ESSS-FDM results in the stochastic drift of the whole population in the direction $m \in \mathbb{R}^N$ determined by the current movement of the population centroid.

The main goal of both strategies is to force population to cross the saddle in the evolutionary landscape between the local extreme currently occupied and the basin of attraction of a better-fitted extreme. Such a policy finally leads to finding at least one of the global extremes.

A detailed description of these strategies together with the test results may be found in the papers [118], [122].

Parameters adaptation using another genetic operations (α.1.3)

We recall once more the individual representation (x, s), $x \in U$, $s \in Q$ introduced at the start of Section 5.3.1. These strategies also posses genetic operations that can process the parameter vectors $s \in Q$, so we define the second genetic algorithm with the genetic universum Q.

Let us consider the execution of the single genetic operation g_i for the fixed index i (see formula 5.1) for the parental individuals $(x^1, s^1), \dots, (x^p, s^p)$. We will perform it in two steps:

1. First we use the proper genetic operation to the sequence of parameters $(s^1, \dots, s^p)$ producing the new parameter vector s'.
2. We execute the operation g_i on the string $(x^1, \dots, x^p, s')$ producing the offspring x'.

A detailed description of these strategies together with the test results may be found in the papers written by Bäck [12], [13], Grefenstette [77] and in the Arabas monograph [5].

5.3.2 Strategies with a variable life time of individuals (α.2)

Adaptation strategies that explicitly determine the life time of individuals lead to the variation of the population's cardinality. This quantity, becomes the new, independent parameter that controls the evolutionary search dynamics. There are two different opinions that explain the influence of this parameter on the genetic search ability:

Michalewicz [110] is of the opinion that a large population performs a better global search because it better fills the admissible domain $\mathcal{D}$ (in particular, the set of phenotypes $\mathcal{D}_r$). Strategies proposed by this author lengthen the individual's life at the initial phase of evolution in order to perfectly fill the admissible domain and to generate the individual with a phenotype close to at least one global extreme of the objective function, or individuals located in the basin of attraction of such an extreme. The individual's life time was shortened in the second phase of evolution in order to strengthen the landscape exploitation (e.g. in order to better concentrate population phenotypes close to the global extreme). Such an idea seems to be effective in the case of bounded, moderately-large admissible domains.

Galar [71] underlines the greater mobility of small populations which results in faster checking of the whole admissible domain. The computational cost of the single evolution step is much lower because of much fewer fitness evaluations. Large populations are the worst suited for passing the evolutionary barriers, staying for a long time in the basin of attraction of the single local extreme. Strategies based on this idea decrease the population size by shortening the individual's life if the evolutionary process is stagnated (e.g. the trap test is satisfied). They seem to be effective in the case of infinite or huge admissible domains that could not be filled satisfactorily by population phenotypes.

Evolutionary strategies with a variable life time of individuals introduce an additional (aside from μ and λ) defining parameter κ which determines the maximum life time of individuals. The specific kind of selection controlled by the imposed individual life time is also utilized (see e.g. [5]):

1. Compute the life time $L(x)$ of the individual with the genotype $x \in U$.
2. If $L(x) > \kappa$ then the individual is removed from the population.
3. The usual type selection is performed among surviving individuals (e.g. proportional selection, see Section 3.4).

The following formulas that allow us to compute the individual life time were quoted after Michalewicz [110], Bäck Fogel and Michalewicz [15], [10], [11] and Arabas [5]. Initially, we introduce three parameters: l_{min}, l_{max}, $l_{av} = \frac{1}{2}(l_{min} + l_{max})$ that denote minimum, maximum and mean individual life time assumed for the whole evolutionary process. We denote moreover:

$$\begin{cases} f_{min}(t) = \min_{\tau=t_s,\dots,t} \{\min_{x \in P_\tau} \{f(x)\}\} \\ f_{max}(t) = \max_{\tau=t_s,\dots,t} \{\max_{x \in P_\tau} \{f(x)\}\} \\ f_{av}(t) = \dfrac{1}{t - t_s} \sum_{\tau=t_s,\dots,t} \left(\dfrac{1}{\lambda(t)} \sum_{x \in P_\tau} f(x) \right) \\ t_s = \max\{0, t - \kappa + 1\} \end{cases} \tag{5.13}$$

where κ is the strategy parameter and $\lambda(t)$ stands for the offspring cardinality in the genetic epoch t.

The first formula suggests the individual life time proportional to its fitness value:

$$L(x) = l_{min} + \frac{l_{max} - l_{min}}{f_{max}(t) - f_{min}(t)} f(x) \tag{5.14}$$

The next two formulas are only slight modifications of the previous one:

$$L(x) = \max\left\{l_{max}, l_{min} + \frac{l_{max} - l_{min}}{f_{max}(t) - f_{min}(t)} f(x)\right\} \tag{5.15}$$

$$L(x) = \begin{cases} l_{min} + \dfrac{l_{av} - l_{min}}{f_{av}(t) - f_{min}(t)} f(x) & \text{if} \quad f(x) < f_{av} \\ l_{av} + \dfrac{l_{max} - l_{av}}{f_{max}(t) - f_{av}(t)} f(x) & \text{if} \quad f(x) \geq f_{av} \end{cases} \tag{5.16}$$

In strategies discussed in this section both the parameters λ and μ become dependent on the genetic epoch counter. Two rules that determine the dependency between the number of parental and offspring individuals can be applied:

a. $\lambda(t) = const.$ and $\mu(t) \leq \mu_{max}(t)$ where $\mu_{max}(t)$ corresponds to the situation in which $L(x) = l_{max}\ \forall x \in P_t$.

b. $\dfrac{\lambda(t)}{\mu(t)} = const.$

Detailed descriptions of the strategies mentioned above, completed by test results, are contained in the papers [4], [5], [164], [7]. These strategies generally increase the intensity of local search in the late phase of evolution, therefore they may be located in class A.

Obuchowicz and Korbicz [118], [120] introduced the strategy *Evolutionary Search with Soft Selection - Varying Population Size* (ESSS-VSP). Each individual x has the initial life time L_0 assigned at its creation. Its life time $L(x)$ varies according to its fitness history.

$$L(x) = \left\lceil l_m \frac{f(x)}{f_m(t)} \right\rceil \tag{5.17}$$

The individual's life time is modified by the ratio between its fitness and the maximum fitness $f_m(t)$ of the whole population P_t at the tth epoch. The quantity l_m stands for the strategy parameter. The population's cardinality grows if the mean population fitness grows, which enforces the local search in the basin of attraction of the single extreme. If the stagnation of the fitness increment is observed i.e. the mean fitness in the population bounds to the maximum one, then the number of individuals drops, which increases the population mobility, then facilitates the passage to the basin of attraction of the better fitted local extreme. This strategy may be classified into group B.

5.3.3 Selection of the operation from the operation set (α.3)

The main idea of this strategy is the personal selection of the genetic operation by each particular operation executed on the individual. The selection is performed from the prescribed set of operations GO. This set may be invariant with respect to the genetic epoch counter or may vary when passing between two consecutive epochs. Practically, we will consider the sequence of operation sets $\{GO^t\}$, $t = 0, 1, 2, \ldots$ in the second case. The sequence element GO^t will be available for the operation selection at the epoch t. The set GO or each set GO^t may posses an internal structure which reflects the appearance of different types of operations. In particular we may have

$$GO = \bigcup_{i \in I_{OT}} GO_i, \quad GO^t = \bigcup_{i \in I_{OT}} GO_i^t \tag{5.18}$$

where I_{OT} denotes the finite set of operation types.

Selection of the operation according to the still probability distribution (α.3.1)

Each genetic operation performed at the tth genetic epoch is preceded by the one time sampling from the set GO according to the invariant probability distribution $p \in \mathcal{M}(GO)$. The sampling is usually multiple, which does not restrict the set GO by consecutive operations in the same epoch. In the case of more than one type of genetic operation, we first sample the operation type i from the set I_{OT} and then the operation using the probability distribution $p_i \in \mathcal{M}(GO_i)$. Both samplings are multiple ones. More information about these adaptation strategy instances may be found in the papers of Davis [52], [53], Julstrom [90], Arabas [5] and Stańczak [181].

Selection of the operation according to the self-adaptive probability distribution (α.3.2)

The first strategy from this group consists of sampling only one genetic operation for the single genetic epoch t. This sampling is performed according to the probability distribution $p^t \in \mathcal{M}(GO)$ which is adopted during the evolution process. In the case of more than one type of genetic operation performed consecutively (e.g. first mutation and next crossover), a single operation for each group is selected at the epoch t. Probability distributions $p_i^t \in \mathcal{M}(GO_i^t)$, $i \in I_{OT}$ used for each operation type selection are also adopted by passing between two neighboring genetic epochs. Each sampling is a multiple one as previously. Let us denote one of the best fitted individuals in the epoch t of the strategy.

$$\hat{x}(t) = \arg\max_{y \in P_t} \{f(y)\} \tag{5.19}$$

We introduce the monitoring function

$$\psi(x, g, t) = \begin{cases} 0 & \text{if } f(x) \leq f(\hat{x}(t)) \\ f(x) - f(\hat{x}(t)) & \text{otherwise} \end{cases} \tag{5.20}$$

where x is the genotype of the individual from the population P_t which was created by the operation g in the epoch t. Then we determine the price of the amount $\alpha\psi(x, g, t)$ which is next distributed among operations that lead to the creation of the individual x during the T last epochs. The positive numbers α and T are the parameters of this strategy. The final price for the particular operation g in the genetic epoch t is computed according to the formula

$$\psi_g(t) = \max \left\{ \psi_{min}, \sum_{\tau=t}^{t-T} \sum_{x \in P_t} \psi(x, g, \tau) \right\} \tag{5.21}$$

where ψ_{min} is some positive constant. We choose some strictly positive probability distribution $p^0 \in \mathcal{M}(GO)$, $p^0(\{g\}) > 0 \ \forall g \in GO$ at the start of the strategy. The modification of the probability distribution is performed by using the formula

$$p^{t+1}(\{g\}) = p^t(\{g\}) - \frac{\psi_g(t)}{\sum_{h \in GO} \psi_h}. \tag{5.22}$$

Quantities $\{p^{t+1}(\{g\})\}$, $g \in GO$ are then normalized so that the following conditions are satisfied

$$\sum_{g \in GO} p^{t+1}(\{g\}) = 1, \ p^{t+1}(\{g\}) > p_{min} \ \forall g \in GO \tag{5.23}$$

where p_{min} is also the constant that characterizes this strategy. The description of the above strategy follows the papers of Davis [52], [53], Julstrom [90], Stańczak [181] and the Arabas monograph [5].

Another kind of modification of the operation sampling probability was proposed by Stańczak [182]. He first introduced the quality coefficients.

$$\{q^t(x, g)\}, \ x \in P_t, \ g \in GO^t \tag{5.24}$$

The probability distribution belonging to $\mathcal{M}(GO^t)$ that will be utilized for the genetic operation sampling will be constructed separately for each genotype $x \in U$. This kind of strategy makes sense only in the case of finite genetic universa $\#U < +\infty$. Let $p^t(x, \cdot) \in \mathcal{M}(GO^t)$ be the probability distribution used for genetic operation sampling which will further act on the genotype x in the genetic epoch t. The suggested modification procedure is the following.

$$p^t(x, \{g\}) = \frac{q^t(x,g)}{\sum_{h \in GO^t} q^t(x,h)}$$

$$q^0(x, \{g\}) = q^0, \; g \in GO^0$$

$$q^{t+1}(x,g) = \begin{cases} q^0 + \dfrac{\Delta^{t+1}(x,g)}{f_{max}(t)} + \alpha q^t(x,g) & g = actual \\ q_{new}(x) & g = new \\ q^t(x,g) & \text{for other operations } g \end{cases} \tag{5.25}$$

where:

$\Delta^{t+1}(x,g)$ positive fitness increment of the individual x resulting from the action of the operation g,
actual the genetic operation currently selected in the $t+1$ epoch,
new the new selected operation in the $t+1$ epoch,
q_0 small number, which guarantees the non-zero probability of sampling,
$\alpha \in (0,1)$ "forgetting" coefficient,
q_{new} arbitrary value of the new operation.

This strategy has been tested for various TSP problem instances (see Stańczak [182]).

All the methods described above prefer genetic operations which force the fast increment of the fitness value. Such strategies belonging to class A. may be successfully applied at the initial phase of evolution. They are particularly advantageous if the specialized operations are included beside the classical ones: crossover and mutation. In such situations the ontogenetic operation selection as in formula 5.25 would be a costly, but effective solution.

Davis [52], [53], Julstrom [90] also described another strategy of the adoptive genetic operation selection. Each genetic operation obtains its own time period (number of genetic epochs) during which it can affect the population individuals. The overall fitness increment which appears in this period may be assigned to this genetic operation. Depending on this fitness increment, the probability distribution of selecting among the current operation and the operations which were active before was established.

Concurrent breeding of the genetic operation population (α.3.3)

The application of evolutionary methods for the genetic operation selection was firstly suggested by Grefenstette [77], Raptis and Tzefastas [132]. The quite similar strategy of concurrently processing individuals that represent potential solutions to the target global optimization problem and the population of genetic operations acting on these individuals was introduced by Stańczak [182]. The single genetic epoch for the operation's population corresponds to $T > 1$ epochs for the solution's population. The evolutionary algorithm of the $(\mu+1)$ type that modifies the operation's population utilizes the following fitness function $f_Q : \left(\bigcup_{t=0,1,\dots} GO^t\right) \to \mathbb{R}_+$ which is changed after each T-length period of genetic epochs for the algorithm that solves the target problem.

$$f_Q^{t+T}(g) = q^0 + \sum_{k=1}^{T} \left(\sum_{x \in P^{t+k}} \frac{\Delta^{t+k}(x,g)}{f_{max}(t+k)} + \alpha q^{t+k+1}(x,g) \right) \tag{5.26}$$

In the above formula the operation g is taken from the set GO^{t+T}. Other notations are identical to notations used in the formula 5.25. The type of evolutionary algorithm $(\mu + 1)$ which deals with operations informs us that only one new operation supplements the parental pool before selection is performed. The evolutionary algorithm that solves the target global optimization problem may be an arbitrary one of the type $(\mu \stackrel{+}{,} \lambda)$.

5.3.4 Introducing local optimization methods to the evolution (α.4)

Some evolutionary strategies introduce the local optimization methods acting in the admissible set $\mathcal{D} \subset \mathbb{R}^N$ into the stochastic search scheme. Their application is restricted, of course, by the proper regularity of the objective function. We will use the encoding operator $code : U \to \mathcal{D}$ and the inverse coding $dcode : \mathcal{D} \longmapsto U$. The mapping code is injective while dcode is surjective, which takes a constant value on some neighborhood of each phenotype from $\mathcal{D}_r \subset \mathcal{D}$. Moreover, the coherency condition holds $\forall\, x \in U \; dcode(code(x)) = x$ (see definition 3.2). We assume that the dcode mapping can be extended in some way to the whole admissible domain $\mathcal{D}$ so that it is the well-defined function $dcode : \mathcal{D} \to U$ (see fourth item in Remark 3.3).

The result of the local optimization method which is activated at the point $y \in \mathcal{D}$ will be denoted as in Section 2.1 by $loc(y) \in \mathcal{D}$. We assume that the local method is "non deteriorative" on the set of phenotypes $\mathcal{D}_r$ which means that

$$\Phi(loc(code(x))) \geq \Phi(code(x)), \quad \forall x \in U. \tag{5.27}$$

The simplest possibility of including the local method in evolutionary computation is monitoring the set of pairs

$$\{dcode(loc(code(x))), \Phi(loc(code(x)))\}, \quad \forall x \in P_t \tag{5.28}$$

and then exploiting monitoring results in various adaptation strategies. Other possibilities will be presented in the next two sections.

Selection by the local method (Baldwin effect) (α.4.1)

This complex computing technology utilizes the local optimization method by fitness evaluation. Basically, the fitness of the individual with the genotype

$x \in U$ is computed as $\Phi(\text{loc}(\text{code}(x)))$ or the simple, monotonic transformation value of this quantity e.g. scaling or non-linear "scaling" $Scale(\Phi(\text{loc}(\text{code}(x))))$ (see Section 3.2).

If the local optimization method is sufficiently accurate, then

$$\Phi(\text{loc}(\text{code}(x))) \approx \Phi(y^+) \tag{5.29}$$

where y^+ is the local extreme to Φ so that $\text{code}(x)$ lies in its set of attraction $x \in R^{loc}_{y^+}$ defined similarly as in Section 2.1 (see definition 2.5). The procedure described above leads to the fitness flattening for all individuals whose phenotypes belong to the set of attraction of the particular local extreme of the function Φ (see Figure 5.5).

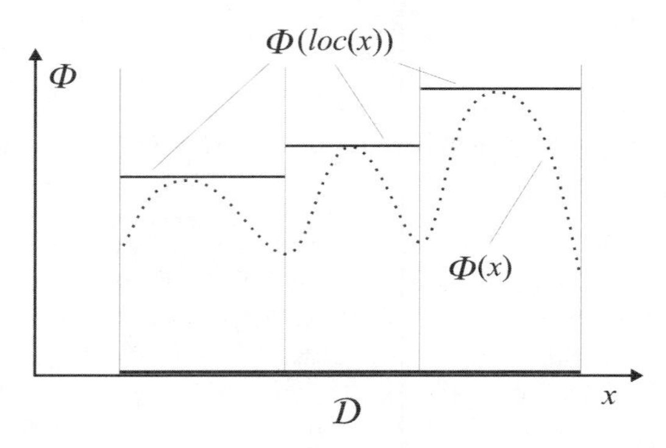

Fig. 5.5. The fitness flattening resulting from the local method application by individual evaluation (Baldwin effect).

The procedure described before may be related to the Baldwin effect observed in biology, which consists of the influence of the acquired features, which is compared here to the activity of local optimization methods which can improve the individual fitness.

This strategy was discussed in the papers of Anderson [2], Whitley, Gordon, Mathias [197] and Arabas [5].

The fitness flattening for individuals whose phenotypes belong to the set of attraction of the particular local extreme of the objective function leads to the uniform distribution of individuals among sets

$$U^{loc}_{y^+} = \text{dcode}(R^{loc}_{y^+} \cap \mathcal{D}_r) \tag{5.30}$$

where y^+ are the local extremes to the objective Φ. As a consequence, the number of individuals whose phenotypes are located close to the set of attraction border $\partial R^{loc}_{y^+}$ is much bigger than in the case of traditional individual evaluation by the function $\Phi(\text{loc}(\text{code}(x)))$ (or by the $Scale(\Phi(\text{loc}(\text{code}(x))))$ function).

If the population phenotypes are mainly enclosed in the set of attraction of the single local extreme R^{loc}_{y+}, then the larger number of individuals which are close to the attractor's border ∂R^{loc}_{y+} may increase the probability of passing the population to the neighboring attractor R^{loc}_{z+} (exactly to the population passage from the set U^{loc}_{y+} to U^{loc}_{z+}). The passage would happen by the mutation of such individuals moving their phenotypes outside the border, to the set R^{loc}_{z+} and automatically evaluated better than at the previous location if $\Phi(z^+) > \Phi(y^+)$ (see formula 5.29). Such an effect appears stronger if the encoding satisfies the Arabas condition 3.3 (see [5], Section 4.2 also see Remark 3.5), because the closeness of the attraction sets R^{loc}_{y+} and R^{loc}_{z+} implicates the closeness of genotype sets U^{loc}_{y+}, U^{loc}_{z+}.

The strategies that utilize local optimization methods by individual evaluation help the population to pass evolutionary barriers, so they may be classified to group B.

Gradient mutation and other genetic operations using local methods (Lamarc effect) (α.4.2)

This strategy consists of introducing the new genetic operations inspired by the local optimization method started for the phenotype of its argument. Perhaps the simplest case of such an operation is the "gradient mutation"

$$U \ni x \to x' = \mathrm{dcode}(\mathrm{loc}(\mathrm{code}(x))) \in U \tag{5.31}$$

This operation was introduced by Littman and Ackley [105]. Burczyński and Orantek [41] utilized the strategy in which only the best fitted individual is transformed by this operation in each genetic epoch. They implemented loc as steepest slope, conjugated gradient and variable metric methods. Burczyński and Orantek [41] also give the test results for the Rastrigin benchmark and the weight optimization by neural network learning. They obtained about 150% speedup in comparision to the simple genetic algorithm when finding the global optimizer.

Smith [174], Yen, Liao, Lee and Rendolph [206] suggest the crossover supported by the single step of the Neldewr-Mead crawling simplex method.

Introducing genetic operations that are based on the local optimization methods quicken the concentration of individual phenotypes close to the local extremes of the objective function Φ, so such strategies may be classified to group A. This group of operations has to be carefully applied (e.g. only in some regions of the admissible set or for a small fraction of population individuals) because the may significantly weaken the global search ability.

5.3.5 Fitness modification (α.5)

This idea allows us to design perhaps the most effective single population genetic algorithms that perform the global search in the whole admissible domain $\mathcal{D}$.

Random fitness modification (α.5.1)

The fitness is modified randomly, according to the formula

$$(f(x))' = f(x) + s\,\mathcal{N}(0,\sigma) \;\; \forall x \in U \tag{5.32}$$

where $s \in \mathbb{R}^+$ stands for the scaling parameter. The normally distributed random variable $\mathcal{N}(0,\sigma)$ is sampled independently by each fitness evaluation. The strategy then has two parameters s, σ. A more extended description of this strategy is given by Arabas [5]. Its application results in smoothing the fitness function graph so that the less distinct local extremes (the basin of attraction has little volume in comparison to the whole volume of the admissible domain, and the fitness variation over this basin is small) do not attract a significant fraction of the population. As a result, this strategy quickens the bounding of individuals to global extremes.

Niching (α.5.2)

This technique, which appears in many variants, consists of forcing the population to spread into several groups of individuals. Separate groups concentrate near separate local extremes of the objective function Φ. This kind of strategy may be classified to group C. This coercion is performed by the proper modification of the fitness function during the evolution process. Basic information about this technology may be found in the Goldberg book [74], Arabas [5] and Chapter written by Mahfoud [107].

Fitness sharing (α.5.2.1)

This method consists of decreasing the evaluation (fitness value) of each individual with the genotype $x \in U$ if its phenotype code(x) is closely surrounded by phenotypes of other individuals. The fitness modification may take the following form

$$(f(x))' = \frac{f(x)}{\sum_{y \in P_t} f_s(d(\mathrm{code}(x), \mathrm{code}(y)))}, \quad \forall x \in P_t \tag{5.33}$$

where d is some metric in the space of phenotypes V, the function $f_s : \mathbb{R}_+ \to [0,1]$ is called "sharing function" so that $f_s(0) = 1$ and $\lim_{\zeta \to +\infty} f(\zeta) = 0$. Sharing functions are usually taken as non-increasing in their whole domain. The typical proposition is the function that restricts the range of fitness sharing effect

$$f_s(d) = \begin{cases} 1 - \left(\dfrac{d}{\sigma_{share}}\right)^{\alpha} & d < \sigma_{share} \\ 0 & \text{otherwise} \end{cases} \tag{5.34}$$

where the real numbers σ_{share} and α stand for parameters of this function, such as sharing range and sharing degree respectively.

Such a fitness adaptation technique, well tuned to the global optimization problem to be solved, forms groups of individuals called "niches" whose phenotypes are concentrated near distinct local extremes of the objective function Φ. Fitness sharing strategies can be assigned to group C. because of their global search ability. The reader is referred to Goldberg and Richardson [73], Goldberg [74], De Jong [54] and Spears [177] for details.

Sequential niching (α.5.2.2)

The fitness modification performed here leads to the leveling of the objective function graph in the central part of the basin of attraction of local extremes already encountered. Selection removes individuals from such areas forcing them to find regions of larger fitness e.g. the basins of attraction of other extremes not yet found. Sequential niching was introduced by Beasley, Bull and Martin [19]. Obuchowicz and Patan [122] verified the fitness modification by using the formula

$$(f(x))' = f_m(t) \exp\left(-\sum_{i=1}^{N}\left(\frac{\text{code}(x)_i - (y_{center}(t))_i}{\sigma_i}\right)^2\right) \tag{5.35}$$

where

$$(y_{center}(t))_i = \frac{1}{\mu}\sum_{x \in P_t} \text{code}(x)_i\,\eta(x)\,, \quad i = 1, \dots N \tag{5.36}$$

are coordinates of the expected centroid of the phenotypes that correspond to the population P_t and

$$\sigma_i^2 = \frac{1}{\mu}\sum_{x \in P_t} (\text{code}(x)_i - m_i)^2\,, \quad i = 1, \dots N \tag{5.37}$$

stands for the variance of the individual location at the ith direction, while $f_m(t)$ is the maximum individual fitness that appears in the population P_t.

Arabas [5] delivered another formula for fitness modification that leads to population niching.

$$\begin{cases} (f(x))' = f(x)\,\text{Deter}(x, x^*) \\ \text{Deter}(x, x^*) = \min\left\{\left(\frac{d(\text{code}(x) - \text{code}(x^*))}{r}\right)^\alpha, 1\right\} \end{cases} \tag{5.38}$$

where x^* is the genotype of one of the best fitted individuals in the population P_t, the real number α is the strategy parameter while r stands for the range of modification. It can be evaluated using the next formula.

$$r = \frac{\sqrt{N}}{2\,\sqrt[N]{p}} \tag{5.39}$$

where p is the predicted number of local extremes.

1: $t \leftarrow 0$;
2: Intialize P_0;
3: $f \leftarrow f_0$;
4: **repeat**
5: Evaluate P_t;
6: Distinguish the best fitted individual x^* from P_t;
7: **if** (*trap_test*) **then**
8: Memorize x^*;
9: $f \leftarrow f'$;
10: **end if**
11: Perform selection with the fitness f;
12: Perform genetic operations;
13: $t \leftarrow t + 1$;
14: **until** (*stop_condition*)

Algorithm 1: Draft of the ESSS-DOF strategy

Obuchowicz [117] exploits the fitness modifications defined above (see formulas 5.35 - 5.39) in his strategy called *Evolutionary Search with Soft Selection - Deterioration of the Objective Function* (ESSS-DOF). The draft of this strategy is presented as the Algorithm 1. Test results of this effective approach to the global search were presented in the papers of Obuchowicz [118], Obuchowicz and Patan [122], Obuchowicz and Korbicz [120].

Trap tests applied here are the same as those described in Section 5.3.1. The logical variable *trap_test* is *true* if

1. the mean fitness in the population increases less than $p\%$ during the last n_{trap} genetic epochs, or
2. the displacement norm of the centroid of the set of phenotypes associated with the population is less than the mutation range σ measured in the space V (exactly the standard deviation of the individual phenotype displacement resulting from mutation) during the last n_{trap} genetic epochs.

The logical variable *stop_condition* is true if the proper stopping rule for the whole strategy is satisfied. The simplest possible stopping rule may be the limited number of genetic epochs after the last fitness modification during which no trap was found.

Telega and Schaefer [184], [185], [186] introduced the sequential niching strategy that utilizes the following fitness modification formula

$$(f(x))' = \begin{cases} f(x) & \text{if} \quad \text{code}(x) \in \mathcal{D} \setminus CL(t) \\ f_{min}(t) & \text{if} \quad \text{code}(x) \in CL(t) \end{cases} \tag{5.40}$$

where $f_{min}(t)$ is the minimum fitness value encountered in the genetic epochs $0, 1, \ldots, t$. If the minimization problem is solved, then $f_{min}(t)$ corresponds to the maximum value of the objective $\Phi_{max}(t)$ so that

$$\Phi_{max}(t) = \max \left\{ \Phi(y) |\, y = \text{code}(x),\, x \in \bigcup_{\tau=0}^{t} P_\tau \right\} \tag{5.41}$$

The set $CL(t) \subset \mathcal{D}$ stands for the sum of the central parts of the basins of attraction of local extremes recognized up to the tth genetic epoch. A detailed description and analysis of this strategy will be made in Section 6.3.

Temporary fitness perturbations – impatience strategy (α.5.3)

This group of strategies allows us to modify the fitness along the evolution process (along the progress of genetic epochs) but they do not lead to population niching. One interesting case of such a strategy, called "impatience strategy", was introduced by Galar and Kopciuch [71]. It utilizes a scheme similar to ESSS-DOF presented by the Algorithm 1. The fitness modification is performed according to the formula

$$\begin{cases} (f(x))' = \alpha(x)\, f(x) \\ \alpha(x) = \dfrac{\beta}{\text{diam}(\text{code}(P_t))} |\text{code}(x) - y_{center}(t)| \\ \text{diam}(\text{code}(P_t)) = \max_{x,y \in P_t} \{ |\text{code}(x) - \text{code}(y)| \} \end{cases} \tag{5.42}$$

where $y_{center}(t)$ is the centroid of the population's P_t phenotypes and β stands for the strategy parameter. If the trap test defined similarly as in case of ESSS-DOF, turns back $trap\_test = false$ then genetic computations are performed without the fitness modification. If $trap\_test = true$ then the fitness modification given by the formula 5.42 results in drawing aside the individual's phenotypes that are concentrated close to the particular local extreme of Φ. The population's phenotypes usually polarize into two groups which are located on the opposite sides of the local extreme currently occupied. These groups go around the extreme, preserving the opposite location mainly due to the crossover symmetry. If at least one of this group finds the saddle in the evolutionary landscape, then approximately half of the population falls into the set of attraction of another extreme (if such a set is located sufficiently close to the saddle).

This strategy may help populations to cross the saddle in order to attain the basin of attraction of the better fitted local extreme. It may be classified then to group B.

5.3.6 Additional replacement of individuals (α.6)

These strategies generally mobilize the population for the periodical or permanent checking of the whole admissible domain $\mathcal{D}$, so they may be classified to group B.

Introducing random individuals (α.6.1)

The offspring O_t (see the scheme in Figure 3.2) is enriched by several individuals produced in this same way as individuals of the initial population P_0, i.e. by sampling with the same probability distribution as by P_0 creation. Such a procedure may be performed permanently (for all $t = 1, 2, \dots$) or periodically in selected genetic epochs. This strategy is discussed in [5].

Periodic population refreshment (α.6.2)

This group of strategies postulates the almost total replacement of the population's individuals by new ones sampled according to the probability distribution, which may depend on the state and the history of the genetic search.

This kind of strategy was initially described by Goldberg [75]. The search performance, either in genotype or phenotype representation, is monitored. If stagnation in evolution is observed, the population is reinitialized by using the cataclysmic mutation (see Eshelman [64]) and the best fitted individual as the pattern string.

Krishnakumar [100] proposed the strategy called Micro-Genetic Algorithm (μGA) that processes small, several-individual populations (typically 5 individuals). If the population diversity sufficiently decreases, then the best fitted individual is kept and the remaining part is re-sampled using a uniform distribution. The μGA strategy does not use mutation.

A similar idea was described in [5] and may be applied to the arbitrary ($\mu \overset{+}{,} \lambda$) evolutionary algorithm. If the trap test of type 1 is satisfied (see Section 5.3.1) then the algorithm is restarted with utilization of the best fitted individual $\hat{x} \in U$ that appears in the history. A new population is created by using the large-range mutation e.g. normal mutation with the mean value $\hat{x}$ and a sufficiently large standard deviation.

Pre-selection and crowding (α.6.3)

Individuals are replaced in the population if a new individual is produced by the genetic operations. The removal of the individual is governed by the following rules (see De Jong [54], Goldberg [74]):

1. The worst fitted individual being one of the parents or the child.
2. The individual whose genotype is nearest to the genotype of the newly-produced one using some metric in U. The removed individual may be selected from the whole population or from its well-specified sub-multiset. This strategy is called *crowding* (see De Jong [54]).

5.3.7 Speciation (α.7)

Speciation strategies enforce concurrent local searches performed by groups of individuals belonging to the single population, which may lead to finding multiple local extremes of the objective function. This strategy may be classified then to group C.

Arbitrary speciation (α.7.1)

This is perhaps the simplest speciation strategy introduced by Goldberg [74]. It may be formalized as the four steps that are repeated consecutively after the initial population P_0 is created.

1. Partition the population P_t for several sub-multisets $P_t^1, \dots, P_t^{Spec}$.
2. Perform genetic operations on each of the sub-multisets $P_t^i,\ i=1,\dots,Spec$ separately.
3. Merge the sub-multisets and perform common selection.
4. Go to step 1 if the stop criterion is not satisfied, otherwise stop.

Restricted range of crossover – mating (α.7.2)

Deb and Goldberg [57] introduced the genetic algorithm called *mating* in which crossover is restricted to individuals whose phenotypes are sufficiently close to one another with respect to the metric in V. Their paper also contains advantageous test results for the multimodal objective optimization.

The first parental individual for crossover x^1 is selected from the whole parental population P_t' (see Figure 3.2) with the uniform probability distribution. The second parent x^2 is searched so that the distance between both parental phenotypes is less than $d(\text{code}(x^1), \text{code}(x^2)) < \sigma_{mate}$ where σ_{mate} stands for the parameter of this strategy. If the operation fails (there are no individuals with the phenotype inside the open ball with the center $\text{code}(x^1)$ and the radius σ_{mate}) then new selection of the first parent x^1 is performed.

If the second procedure fails after several samplings of the first parental individual e.g. the phenotype density is less than $(\sigma_{mate})^{-1}$ then both parents are chosen from P_t' with the uniform probability distribution.

Cellular genetic algorithm (α.7.3)

The strategies described in Section 5.3.7 postulate the selection of the second parental individual x^2 to the crossover operation from the neighborhood of the first parental individual x^1. The neighborhood was taken according to the topology, which changes during evolution (changes from one genetic epoch to the next, consecutive epoch). In the cellular genetic algorithm, individuals have stiff topological connections e.g. they are located in nodes of the p-dimensional lattice. These connections are not usually imposed by the topology in the genetic universum U.

For each arbitrary individual $x \in P_t$ its neighborhood $\mathcal{N}_r(x)$ is established. Usually $\mathcal{N}_r(x)$ overlaps the neighborhoods of other individuals in the population forming cells of this strategy. The parameter r denotes the diameter of the neighborhood expressed in units that characterize the particular individual's structure (e.g. it may be the length of the path in the connection graph of the structure). Genetic operations may be performed only for arguments coming from the single cell, so the individual x can be crossed only with other individuals from its neighborhood $\mathcal{N}_r(x)$. The child-individual compete for the place in its cell.

Operations in cells can be performed asynchronously, concurrently for non-overlapping cells. The strategy can be easily implemented in the environment with many processors and local memory e.g. the distributed environment of a computer network. Details of the cellular genetic strategy may be found in the papers of De Jong, Sarma [55], and Seredyński [166].

5.3.8 Variable accuracy searches (α.8)

One of the main disadvantages of the classical genetic algorithms applied for solving continuous global optimization problems is their low accuracy, measured in the space of solutions V (e.g. the space which contain the population's phenotypes). In the case of discrete encoding (e.g. binary affine encoding) the accuracy is restricted by the discretization error imposed by the density and the topology of phenotype mesh $\mathcal{D}_r$ selection.

In both discrete and continuous encoding cases, the low accuracy may result from the low efficiency of the genetic search caused by the small number of individuals in comparison to the volume of the admissible set $\mathcal{D}$ and low progress of evolution slowed down by the necessary diversity improvements.

The natural way to avoid such problems is to modify the classical genetic search techniques toward concentrating the search in the regions of $\mathcal{D}$ in which the solution's appearance probability grows along the evolution. Two single population strategies of this type will be presented in following sections. Two multi-deme strategies which can also modify the search accuracy (HGS and iGP) will be mentioned in Sections 5.4.3 and 5.4.4.

Dynamic Parameter Encoding (DPE) (α.8.1)

This strategy, introduced by Schraudolph and Belew [161], is based on binary, regular affine encoding (see Section 3.1.1) which produces the regular N-dimensional mesh $\mathcal{D}_r \subset \mathcal{D} \subset \mathbb{R}^N$ in the admissible domain. This mesh limits the maximum accuracy which may be obtained in the DPE search. We denote by s the length of binary strings that are used to encode each coordinate of points from $\mathcal{D}_r$, so the total length of strings is $l = N \cdot s$. Such strings form the basic genetic universum Ω.

However, the searching DPE population operates on strings from the set Ω_{DPE} of the total length $l_{DPE} = N \cdot s_{DPE}$, which is significantly shorter than l. At the start of the DPE adaptation process, strings from Ω_{DPE} represent the s_{DPE} most significant bits of genotypes from Ω that encode parameter intervals containing $(s - s_{DPE})$ points at each dimension, so each genotype from Ω_{DPE} corresponds to the brick in $\mathcal{D}$ that contains $N(s - s_{DPE})$ points from $\mathcal{D}_r$. Each string from Ω_{DPE} is completed by a $(s - s_{DPE})$-length suffix randomly generated (with respect to the uniform probability distribution) in order to evaluate its fitness which is originally defined on Ω. Suffix bits are not involved in genetic operations on Ω_{DPE} genotypes, but they are rewritten to the child suffix.

After stagnation in the evolutionary process is observed, the most promising bricks associated with the fixed s_{step} prefix are distinguished. They are called *target bricks*. The search process is focused on target bricks only in the next DPE adaptation steps. In the next step Ω_{DPE} genotypes represent the $\mathcal{D}_r$ points lying in target bricks. They are completed now by the fixed s_{step} prefix and the randomly selected $(s - s_{DPE} - s_{step})$-length suffix in order to evaluate their fitness. Individuals whose phenotypes are contained in target bricks remain while those outside are transformed in some way in order to fall again into target bricks.

The adaptation procedure is repeated until the satisfactory accuracy measured in the phenotype space is obtained or $(s - s_{DPE} - n \cdot s_{step})$ becomes less than zero for the nth adaptation step (searching population intervals represent the less significant bits of genotypes from Ω).

The strategy may significantly improve the efficiency of finding the final, global extreme with the maximum accuracy imposed by the network $\mathcal{D}_r$ in comparison with the traditional genetic search performed by the population of individuals with genotypes from Ω, because the cascade of short time searches with growing accuracy has a much lower computational cost than a single search with a large population performed during many more genetic epochs. However, the strategy may fail to find the global extreme if bad target bricks are selected at the initial steps.

Delta Coding (α.8.2)

Some similar ideas to the consecutive increasing of the search accuracy as in Dynamic Parameter Encoding are represented in the Delta Coding strategy, invented and studied by Whitley, Mathias and Fitzhorn [199]. The strategy is also related to the discrete, affine, binary encoding code : $\Omega \rightarrow \mathcal{D}_r$ which maps genotypes of the length $l = N \cdot s$ from the genetic universum Ω to the adequate set of phenotypes $\mathcal{D}_r \subset \mathcal{D} \subset \mathbb{R}^N$ (see Section 3.1.1).

The whole strategy is composed of two phases. The first one is the conventional genetic computation (authors of the paper [199] used the effective GENITOR algorithm with elitist selection) in which Ω and $\mathcal{D}_r$ play the roles of genetic universum and phenotype set respectively. This phase is finished if the population is sufficiently concentrated (i.e. the population diameter in the Hamming metric is sufficiently small), which indicates stagnation in the evolution process. Additionally, the best fitted individual $\hat{x} \in \Omega$ is distinguished and memorized at the end of this phase.

Next, the strategy passes to the second phase called the *delta phase*. Now the search is performed in the restricted area neighboring the best fitted individual $\hat{x}$. In order to perform such a delta genetic search, the new genetic universum Ω_δ which gathers the binary strings of the length $l_\delta = N(1 + s_\delta)$ is defined. A string $\Delta \in \Omega_\delta$ encodes each coordinate increment at s_δ bits plus one bit for encoding the coordinate increment sign. Of course s_δ has to be much less than s. In order to evaluate individuals from Ω_δ the new fitness $f_\delta : \Omega_\delta \ni \Delta \rightarrow f(\hat{x} + \Delta) \in \mathbb{R}_+$ is established. The sum $(\hat{x} + \Delta) \in \Omega$ is interpreted as the catenation of the N sums evaluated independently for each s-length substring of $\hat{x}$ and the adequate $(1 + s_\delta)$-length substring of Δ respecting its sign coded by the first bit. No mutation is utilized in the delta phase.

The delta phase is repeated iteratively. The delta population is completely reinitialized at the start to each step. The evolution in each delta step is finished if the diversity in the population sufficiently decreases, i.e. if the Hamming distance between the best and the worst fitted individuals is less than or equal one. The best individual $\hat{x}$ is updated and the range of the delta search is decreased by reducing s_δ after each iteration step. Iterations are finished if the satisfactory accuracy is obtained or the s_δ is reduced to zero.

From similar reasons to DPE, the delta coding strategy may improve the efficiency of finding a global extreme with the maximum accuracy imposed by the network $\mathcal{D}_r$ in comparison with the traditional genetic search performed by the population of individuals with genotypes from Ω. The special procedure performed in order to preserve the ability of the global search decreases the risk of the delta population getting bogged down in the basin of attraction of the local extreme.

5.4 Multi-deme strategies (β)

The genetic strategies mentioned in the section's title constitute serious competition with single-population ones, especially when solving heavy multimodal global optimization problems. Multi-deme genetic strategies process many equivalent or hierarchically-linked populations called *demes*. They may be compared to a single colony of many species that develop in the common environment. It is worth mentioning that multi-deme strategies, in spite of their formal complication, are significantly less computationally complex (they need much fewer evaluations of the fitness function) in comparison to the single-population ones when solving the hardest multimodal problems, in which the basins of attractions of local extremes are separated by huge areas on which fitness exhibits almost "flat" behavior (plateaus).

Multi-deme strategies are also best suited for implementation dedicated to the multi-processor environment. There is a possibility to distinguish the coarse- or medium-size-grain computational tasks which can be effectively processed in parallel in a distributed environment.

5.4.1 Metaevolution (β.1)

The metaevolution strategy is the parallel process of refining the target genetic algorithm that solves the target global optimization problem in the admissible domain $\mathcal{D}$. The main idea is to use the so-called genetic meta-algorithm to transform the population of the genetic algorithms that solve the target problem. According to the convention introduced in Section 3.9 they are strategies of the type $\left[\mu' \stackrel{+}{,} \lambda' \left(\mu \stackrel{+}{,} \lambda\right)^{\gamma}\right]^{\gamma'}$. These strategies operate then on two levels:

Meta level population $\tilde{P} = (\tilde{U}, \tilde{\eta})$ represents the multiset of genetic algorithms. In particular, each genotype $\tilde{s} \in \tilde{U}$ encodes the genetic algorithm $g_{\tilde{s}}$ which operates on the target level. Meta-genetic operations transform $\tilde{P}$ to the new population of algorithms $\tilde{P}'$ in each genetic epoch on the meta level. It is convenient to denote the meta level population in the set-like form $\tilde{P} = \langle \tilde{s}_1, \ldots, \tilde{s}_{\mu'} \rangle$ imposed by Remark 2.15. The fitness function value $\tilde{f}(\tilde{s})$ on this level is computed as the statistics of the $g_{\tilde{s}}$ behavior on the target level. The arbitrary strategy may be used on this level, but strategies preserving the population size $\mu' = const.$ are preferred for technical reasons.

Target level is composed of μ' genetic algorithms $\{g_{\tilde{s}}\}$, $\tilde{s} \in \tilde{P}$ which operate on separate target populations $P^{\tilde{s}_1}, \ldots, P^{\tilde{s}_{\mu'}}$. The phenotypes of individuals from all the target populations are encoded points from the admissible domain $\mathcal{D}$ of the target global optimization problem. At each genetic epoch of the meta level $\gamma > 1$ genetic epochs are performed for all target algorithms $\{g_{\tilde{s}}\}$, $\tilde{s} \in \tilde{P}$ in parallel. After running the γ epoch the special non-negative statistics $\tilde{f}(\tilde{s})$ are evaluated for each algorithm $g_{\tilde{s}}$, $\tilde{s} \in \tilde{P}$. These statistics may be the best fitness that occurs in $P^{\tilde{s}}$ or the mean fitness increment in $P^{\tilde{s}}$ during the last γ target epochs. The value $\tilde{f}(\tilde{s})$ will be the fitness of the individual $\tilde{s}$ at the meta level.

Metaevolution may be handled as the generalization of the adaptive strategy described in Section 5.3.1, (α.1.3). The main differences among these two strategies may be reported as follows:

- In the adaptation technique described in (α.1.3) only various versions of this same genetic operation may compete. They may differ in parameter values (e.g. mutation rate). The selection of the proper version of the operation is made, respecting their results obtained on the single individual.

- In the metaevolution strategy, whole genetic algorithms (various strategies, various sets of operations) may compete. They are evaluated based on the statistics that were computed for whole separate target populations.

A more exhaustive description of metaevolution strategies may be found in Freisleben [69]. An interesting strategy, close to the metaevolution idea, is presented by Stańczak [182]. Metaevolution strategies may accomplish various goals A., B. or C. according to assumed statistics $\tilde{f}(\tilde{s})$ and particular meta and target strategies.

5.4.2 Island models (β.2)

The island genetic strategy consists of concurrently processing the fixed number of populations $P_t^1, \dots, P_t^{\mu'}$, $t = 0, 1, \dots$ called "islands". This processing is not necessarily synchronous. All islands are engaged in solving the same global optimization problem in the admissible domain $\mathcal{D}$. The genetic universum, encoding and fitness have to be the same for all island genetic algorithms while these algorithms are not necessarily identical. Additionally, the starting states of island populations $P_0^1, \dots, P_0^{\mu'}$ may differ from one another. The island model of genetic computations may be then classified as $\left[\mu' \overset{+}{,} \lambda' \left(\mu \overset{+}{,} \lambda\right)^{\gamma}\right]^{\gamma'}$.

However, the island model is not the simple redoubling of one or several types of single-population genetic searches. The genetic material is exchanged among islands by migration of small groups of individuals. It is possible because of the unified form of individuals (this same genetic universum and encoding assumed). Most frequently, clones of better fitted individuals are sent to another island by the home island process. The migration may be performed synchronously, according to the fixed topology of islands (e.g. ring topology) or asynchronously with the random selection of the destination island.

Whitley and Gordon [198] applied the Simple Genetic Algorithm in each island and the synchronous migration of the single clone of the best fitted individual, which follows the ring topology of islands. Each island is independently initialized according to the uniform probability distribution on the genotype universum Ω. Obviously it does not imply the equality of all the starting islands. Whitley, Soraya and Heckerdorn [200] explained the mechanism of the island genetic search based on the Markov model applied for each island. The papers [198] and [200] are also rich in examples that illustrate the possibility of occupying the basins of attractions of various local maxims of the linearly separated objective function by separate genetic islands. They presented, moreover, another genetic islands model in which the more effective GENITOR algorithm is implemented at each island. New papers studying island model efficiency were published by Skolicki and De Jong (see e.g. [169], [170]). A review of genetic island models may also be found in the papers of Seredyński [166], Martin, Lieinig and Cohon [109].

Almost all island models may be classified to group C. because of their ability to concurrently search in the domain $\mathcal{D}$ and, as a result, quickly find many local extremes of the objective function.

Because of the coarse computational grain (single island defines the computational task that does not need to communicate intensively with other islands) island strategies perfectly fit the parallel, distributed computation paradigm.

5.4.3 Hierarchic Genetic Strategies (β.3)

Hierarchic genetic strategies were introduced by Kołodziej and Schaefer [157]. The first implementation called HGS, using SGA instances as the genetic engine were described in detail in [96]. The next implementation called HGS-RN, in which the SGA was replaced by the simple evolutionary mechanism with phenotypic encoding, was delivered by Wierzba, Semczuk and Schaefer [201]. The brief description of both implementations presented below mainly follows the last cited paper.

The main idea of the Hierarchical Genetic Strategy is running a set of dependent evolutionary processes in parallel. The dependency relation has a tree structure with a restricted number of levels m. The processes of lower order (close to the root of the structure) represent a chaotic search with low accuracy. They detect the promising regions of the optimization landscape, in which more accurate processes of higher order are activated. Populations evolving in different processes can contain individuals which represent the solution (the phenotype) with different precision. This precision can be achieved by binary genotypes of different length, in the case of binary implementation or by different phenotype scaling, in the case of HGS-RN.

The strategy starts with the process of the lowest order 1 called the root. After a fixed number of evolution epochs the best adapted individual is selected. We call this procedure a *metaepoch* of the fixed period. After every metaepoch a new process of the order 2 can be activated. This procedure is called *sprouting operation*. Sprouting can be generalized in some way to branches of a population's tree of a higher order up to $m-1$. Sprouting is performed conditionally, according to the outcome of the *branch comparison operation*. Details of both operations depend strongly upon the strategy implementation.

Binary implementation HGS

Let us concentrate first on the binary implementation HGS. The HGS genetic process is of the order $j \in \{1, \ldots, m\}$ if the individuals from the evolving population have genotypes of the the length $s_j \in \mathbb{N}$. The lengths of binary strings used in various order processes satisfy the inequality $1 < s_1 <, \ldots, < s_m < +\infty$. The initial population for the new sprouted branch of the order $j+1 \leq m$ contains individuals with prefixes identical to the genotype of the best adapted individual in the process of the order j. Suffixes of the length $s_{j+1}-s_j$ of these individuals are initialized randomly (according to the uniform distribution).

The branch comparison operation in HGS is based on the *prefix comparison*. The operator acts on populations evolving in the processes of two consecutive orders j and $j+1$. Let us assume that we distinguish the best fitted individual x from the branch of the jth order after some metaepochs. If there is at least one individual with the prefix of the length s_j identical to x among $j+1$ order branches, then a new process of the order $j+1$ is not activated.

The special kind of *hierarchical nested encoding* is used in order to obtain the search coherency for branches of various degrees. Let us denote by Ω_s the genetic universum composed of binary codes of the length $s > 0$, so $\Omega_{s_1}, \ldots, \Omega_{s_m}$ stand for the binary genetic universa of branches of degrees $1, \ldots, m$. Each universum is linearly ordered by the relation induced by the natural order among integers represented by binary strings. Moreover, for $j = 2, \ldots, m$ we can represent genetic spaces Ω_{s_j} in the following way:

$$\Omega_{s_j} = \left\{(\omega, \xi),\ \omega \in \Omega_{s_{j-1}},\ \xi \in \Omega_{s_j - s_{j-1}}\right\}. \tag{5.43}$$

First, we define the hierarchical nested encoding for $\mathcal{D} \subset \mathbb{R}$ and then generalize the construction to $\mathcal{D} \subset \mathbb{R}^N$, $N > 1$. We intend to define a sequence of meshes $\mathcal{D}_{r_1}, \ldots, \mathcal{D}_{r_m} \subset \mathcal{D} \subset \mathbb{R}$ so that $\#\Omega_{s_j} = \#\mathcal{D}_{r_j}$, $j = 1, \ldots, m$ and a sequence of one-to-one encoding mappings $\text{code}_j : \Omega_{s_j} \to \mathcal{D}_{r_j}$. We will do it recursively. First, we arbitrarily define the densest mesh $\mathcal{D}_{r_m}$ in $\mathcal{D} \subset \mathbb{R}$ and the encoding $\text{code}_m : \Omega_{s_m} \to \mathcal{D}_{r_m}$ as a strictly increasing function. Next, we arbitrarily define the set of selections $\phi_j : \Omega_{s_j} \to \Omega_{s_{j+1}-s_j}$, $j = 1, \ldots, m-1$ which play a fundamental role in the construction of meshes $\mathcal{D}_{r_j}$, $j = 1, \ldots, m-1$. Finally, we put

$$\mathcal{D}_{r_j} = \left\{\text{code}_{j+1}(\omega, \phi_j(\omega)),\ \omega \in \Omega_{s_j}\right\}. \tag{5.44}$$

Figure 5.6 below shows the sample meshes D_{r_1}, D_{r_2} and $D_{r_3} \subset \mathbb{R}$ in the case of $s_1 = 2$, $s_2 = 3$, $s_3 = 5$, $\phi_1(00) = \phi_1(01) = 1$, $\phi_1(10) = \phi_1(11) = 0$, $\phi_2 \equiv 01$. j = 1, ..., m-1.

In the multidimensional case $\mathcal{D} \subset \mathbb{R}^N$, $N > 1$ we assume the sequence of sub-string lengths $1 < \hat{s}_1 <, \ldots, < \hat{s}_m < +\infty$ used for encoding the single coordinate of phenotypes at each order branch. We have then $s_j = N\hat{s}_j$, $j = 1, \ldots, m$. Next, we define the arbitrary sets $\mathcal{D}^1_{r_m}, \ldots, \mathcal{D}^N_{r_m}$ so that $\mathcal{D}_{r_m} = \mathcal{D}^1_{r_m} \times, \ldots, \times \mathcal{D}^N_{r_m} \subset \mathcal{D}$ and then strictly increasing mappings $\text{code}^i_m : \Omega_{\hat{s}_m} \to \mathcal{D}^i_{r_m}$, $i = 1, \ldots, N$.

As in the one dimensional case the construction of the coarse meshes is based on the selections $\phi_j : \Omega_{\hat{s}_j} \to \Omega_{\hat{s}_{j+1}-\hat{s}_j}$, $j = 1, \ldots, m-1$. Finally, we can set

$$\mathcal{D}^i_{r_j} = \left\{\text{code}^i_{j+1}(\omega, \phi_j(\omega)),\ \omega \in \Omega_{\hat{s}_j}\right\},\ i = 1, \ldots, N,\ j = 1, \ldots, m-1 \tag{5.45}$$

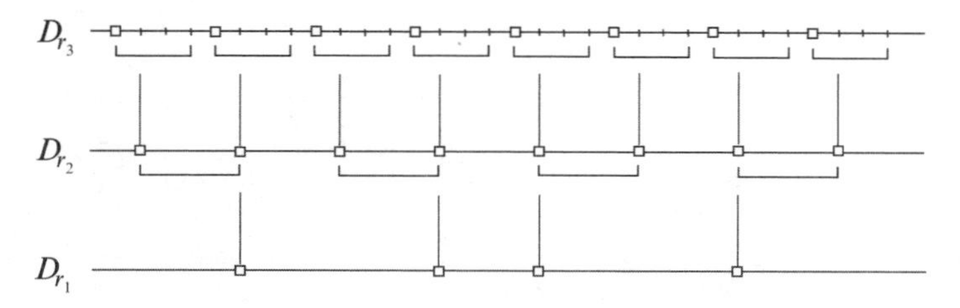

Fig. 5.6. One dimensional nested meshes ($\mathcal{D} \subset \mathbb{R}$) for Hierarchical Genetic Strategy in the case $s_1 = 2, s_2 = 3, s_3 = 5$.

and then

$$\mathcal{D}_{r_j} = \mathcal{D}^1_{r_j} \times, \dots, \times \mathcal{D}^N_{r_j},\ j = 1, \dots, m-1. \tag{5.46}$$

Encoding mappings $\text{code}_j : \Omega_{s_j} \to \mathcal{D}_{r_j},\ j = 1, \dots, m$ are defined as the composition of mappings $\text{code}^1_j, \dots, \text{code}^N_j$ while each code^i_j is taken for the ith substring of the length $\hat{s}_j$ in the argument string $x \in \Omega_{s_j}$.

The Simple Genetic Algorithm defined in Section 3.6 is used for each branch processing during metaepochs. Usually, larger population cardinality is set for lower order branches (close to the root) and much smaller cardinality is set for higher order branches and leafs. Additionally, the mutation rate p_m is set higher for the root and main branches in order to strengthen the wide exploration of the admissible domain.

The well-asymptotic properties of HGS can be mathematically proved (see [96] and [157]). Its high accuracy and the low computational cost, in comparison with island genetic strategies and with the specialized ESSS-SVA algorithms (see Sections 5.4.2, 5.3.1), were experimentally shown for multimodal benchmarks (see [96], [156]). Moreover, its applicability to the difficult engineering problem of minimizing the coordinate measuring machine geometrical errors was presented in [98].

Real encoding implementation HGS-RN

In the HGS-RN both genotypes and phenotypes appearing in each branch of arbitrary order are the N-dimensional vectors of real entries. The genotype universa for branches of different orders are obtained by the proper scaling. Let's assume that the admissible domain is now $\mathcal{D} = [a, b]^N \subset \mathbb{R}^N$, $N > 1$. We introduce the sequence of scaling coefficients $+\infty > \xi_1 >, \dots, > \xi_m = 1$. The genetic universa of the various order branches are defined as (see Figure 5.7)

$$U_j = \left[0, \frac{b-a}{\xi_j}\right]^N,\ j = 1, \dots, m \tag{5.47}$$

thus, encoding functions are given by

$$\text{code}_j : U_j \ni \{x^i\} \to \{\xi_j\, x^i + a\} \in \mathcal{D},\ j = 1, \dots, m. \tag{5.48}$$

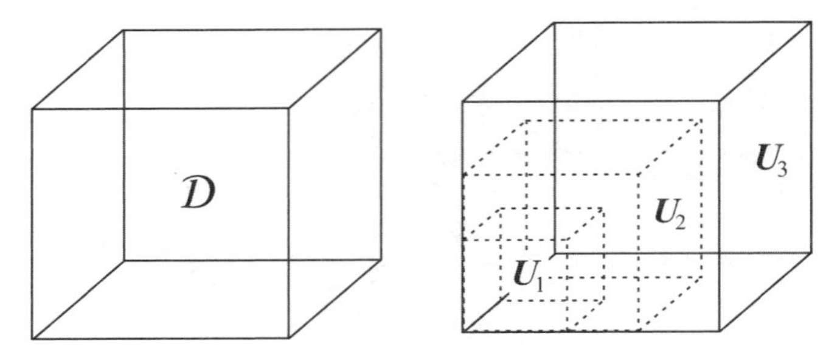

Fig. 5.7. Genetic universa in the HGS-RN strategy

We define moreover, the re-scaling function

$$\mathrm{scale}_{i,j} : U_i \ni \{x^k\} \to \left\{ \frac{\xi_i}{\xi_j} x^k \right\} \in U_j, \; i,j = 1,\ldots,m. \tag{5.49}$$

The initial population for the newly sprouted branch of degree $j+1$ is randomly chosen according to the N-dimensional probability distribution

$$\mathcal{N}((\mathrm{scale}_{j,j+1}(y))^1, \sigma_j), \ldots, \mathcal{N}((\mathrm{scale}_{j,j+1}(y))^N, \sigma_j) \tag{5.50}$$

where y is the best adapted individual in the parental process of the order $j = 1,\ldots,m-1$, is the standard deviation specific for the branch order j.

The branch comparison operation is based on the arithmetic average of genotypes in a population. Let P be a population of the order $j = 1,\ldots,m-1$ and y its best fitted individual currently distinct from P after a metaepoch. Sprouting is not activated if there exists a population P' of the level $j+1$ which satisfies the following condition

$$d(\bar{y}, \mathrm{scale}_{j,j+1}(y)) < c_{j+1} \tag{5.51}$$

where $\bar{y}$ is the genotype's average in the population P', c_{j+1} is the branch comparison constant for $j+1$ order branches and d stands for the Euclidean distance in $\mathbb{R}^N$.

The simple evolutionary techniques based on the normal phenotypic mutation and arithmetic crossover described in Section 3.7 are utilized in all branches. Moreover, the proper repairing operations are used in the case of individuals generated outside the genotype universum. The searches performed by branches of the lower order are less accurate and broader because of the restriction of the genotype set imposed by scaling (lower order branches search in smaller genotype sets). The search in higher order branches may be additionally narrowed by reducing standard deviations in the mutation operation. As in the case of HGS strategy, the population cardinality decreases along the growing degree of branches.

Basic implementation of both HGS and HGS-RS assume synchronization of all branches after each metaepoch while the branches are processed in parallel between these checkpoints. At each checkpoint, branch comparison and sprouting is performed. This partially synchronous approach was described in detail in [156]. A more relaxed approach in which HGS-RN was run as the dynamic collection of intelligent agents was proposed and tested in [113]. There is no total synchronization among computing branches. Instead of the branch comparison and the conditional sprouting mechanisms, special type agents visit populations of the same order reducing the search redundancy.

The HGS-RN strategy has been intensively tested and exhibits exceptional accuracy especially in the case of difficult multimodal benchmarks (see [201]) much higher than HGS.

5.4.4 Inductive Genetic Programming (iGP) (β.4)

This strategy introduced by Slavov, Nikolaev [171], [172] is based on the decomposition of the evolutionary landscape suggested by Stadler [180]. The fitness function f of the complicated behavior is presented as the composition of the set $\{f_n\}$ of simpler mappings. Each of f_n pointed out one of the characteristics of the f variability. One of the effective methods for obtaining the set $\{f_n\}$ is to perform the Fourier transformation of fitness f in the standard orthogonal basis $\{e_i\}$ which is proper for the set of genotypes U.

$$f = \sum_{i=1}^{+\infty} a_i\, e_i, \quad f_i = a_i\, e_i, \; i = 1, \ldots, +\infty \tag{5.52}$$

The strategy is composed of the superior population that explores the landscape imposed by the function f and the finite set of secondary populations, which explore landscapes associated with functions $\{f_i\}$, $i = 1, \ldots\, m$. Individuals from secondary populations may migrate to the superior one. The general rule of such migration is the following:

1. The best fitted individual $\hat{x}$ from the secondary population that evolves with the fitness f_i is selected after some genetic epochs.
2. The individual $\hat{x}$ is transformed using the strictly ascent random walk (see Section 2.2) using mutation as the operation for obtaining the next step sample and the superior fitness f for the sample evaluation.
3. If the random walk described above stagnates in the neighborhood of the local extreme of f, then the process is stopped and its final result is passed to the superior population.

The idea of using secondary populations is to more easily detect the local extremes that are better exposed by the particular secondary fitnesses f_i. In particular, if $f_i = a_i\, e_i$ is the Fourier component of f, then it better exposes local extremes that appear with the specific frequency, imposed by the basic function e_i. The rough location of one such extreme, represented by $\hat{x}$, is then refined by the random walk (see step 2. of this strategy) with respect to the superior fitness f.

The iGP strategy was successfully applied to the problems with complicated fitness functions in the Slavov and Nikolaev papers [171], [172], already mentioned as well as in the paper of Nikolaev and Hitoshi [115].

6

Two-phase stochastic global optimization strategies

Written by Henryk Telega

6.1 Overview of two-phase stochastic global strategies

Many stochastic strategies in global optimization consist of two phases: the global phase and the local phase. During the global phase, random points are drawn from the domain of searches $\mathcal{D}$ according to a certain, often uniform, distribution. Then, the objective function Φ (or $\tilde{\Phi}$, see Section 2.1 Definition 2.1 and Remark 2.2) is evaluated in these points. During the local phase the set of drawn points (the sample) is transformed by means of local optimization methods. The goal of this phase is to obtain an approximation of the global extreme or local extremes.

The two-phase methods can be described as methods for searching maxima or minima (maxima are considered in the monograph [86], minima are considered for instance by Rinnooy Kan, Timmer in [137] and [138]). In Section 6.1 only minimization problems are considered (see Chapter 2 problems $\{\tilde{\mathbf{\Pi}}_i, \tilde{\mathbf{\Pi}}_i^{\mathbf{a}j}\}, i = 1, 2, 3, 4; j = 1, 2, 3$). We limit our considerations to cases in which one can find equivalent maximization problems with non-negative objective functions (see also Chapter 2, Remarks 2.2 and 2.3).

6.1.1 Global phase

Two desirable properties of the global phase are asymptotic probabilistic correctness and asymptotic probabilistic guarantee of finding all local minimizers or maximizers (see definitions 2.16, 2.17).

In this section we assume that random points are cumulated and remembered. The sample stands here for the set of all points generated in subsequent steps of an algorithm from its beginning to a certain point in time. In order to differentiate this sample from the sample defined in Chapter 2 (see schema in Figure 2.2) we could call the former one the cumulated sample. However, for the sake of simplicity, we will call it just sample.

R. Schaefer: *Foundation of Global Genetic Optimization*, Studies in Computational Intelligence (SCI) **74**, 153–197 (2007)
www.springerlink.com

Basic probabilistic space

Let the drawing (generating) of a point from the domain $\mathcal{D}$ be the elementary event. By Ω let's denote the set of elementary events, Σ_Ω will stand for the σ-algebra of subsets of Ω, by Σ we denote the σ-algebra of subsets of $\mathbb{R}^N$. In this chapter random variables and random vectors will be underlined (e.g. $\underline{\Theta}$), their realizations will be denoted without the underline (e.g. Θ). Let $\underline{x_i} : \Omega \rightarrow \mathbb{R}^N$ for $i = 1, 2, \ldots$ be the random vector which maps the random event of choosing a point $x \in \mathcal{D}$ in the ith sampling to this point. In many methods the random points are generated according to the uniform distribution. For such methods $\forall \omega \in \Omega$, $\forall A \subset \Omega$ we evaluate the probability (see [28] Chapter 1)

$$\Pr\{\underline{x_i}(\omega) \in A\} = \frac{\text{meas}(A)}{\text{meas}(\mathcal{D})}. \tag{6.1}$$

Probability of drawing a point from the set $A \in \Sigma$ in the global phase

When we assume the uniform distribution, then the probability of drawing a point from any measurable set $A \subset \mathcal{D}$ after m independent drawings, according to the Bernoulli schema, is equal to

$$1 - \left(1 - \left(\frac{\text{meas}(A)}{\text{meas}(\mathcal{D})}\right)\right)^m. \tag{6.2}$$

Any assumption about the objective function Φ or $\tilde{\Phi}$, which guarantees that the measure of appropriate sets A_ε (see Chapter 2, Section 2.1) for maximization and minimization problems is positive, implies that the probability of finding a point from any of these sets becomes close to 1 when m increases (see [83] page 831, and also papers cited there: Brooks [37], Devroye [59]).

An important question that should be answered is how to generate random points. This problem will not be covered in this book in detail. Apart from using standard generators, for instance, included in various libraries or attached to compilers, other methods can also be used, such as Halton sequences (see Shaw [167]) or using pre-samples and discarding points that are located near points generated and accepted earlier (see Törn, Viitanen [190]).

6.1.2 Global phase - why stochastic methods?

A comparison of the global phase in which points are generated according to the uniform distribution with deterministic strategies like Grid Search (values of the objective function are calculated in points of a regular grid defined on the admissible set) is described by Sukharev [183], Anderssen and Bloomfield [3], Archetti and Betrò [9]. Pure Random Search (see sections 2.2 and 6.1.4) which contains only the global phase has been analyzed there. Analyses show that random points cover the admissible set more efficiently (according to some criteria which take into account the distance from the global minimizer to the nearest point of the grid, and to the nearest point from the random sample) than points of the regular grid, at least if the dimension of the problem is not too low (see also Rinnooy Kan, Timmer [137]).

Sobol in [175] noticed that the projection of points from the uniformly distributed random sample onto any subspace gives the uniform distribution. In the case of a regular grid, such a projection can give groups of points which do not cover the admissible set well (at least for certain global optimization problems). This indicates that, for some problems, stochastic random samples are better than regular grids.

Another argument which works to the advantage of stochastic methods is the fact that in many optimization methods there is no need to know, or assume, in advance the number of points that have to be generated. Sampling can be finished after an optimal or a suboptimal stopping rule (see Section 6.2) is satisfied. Such rules can be defined for stochastic methods. In many methods, after a stopping rule is satisfied, the distribution of points in $\mathcal{D}$ is uniform and the whole set is covered by these points. Such property does not hold for methods based on regular grids.

There are stopping rules which determine in advance the number of points m that have to be drawn (see Section 6.2). However, for certain admissible sets $\mathcal{D}$ it is not easy to define for all m what a regular grid is. We should note here that there are also deterministic strategies that cover uniformly finite closed sets. Sobol in [175] describes the use of infinite sequences (called $LP\tau$ sequences), for which finite subsequences have the so-called measure of non-uniformity less than regular grids. However, these issues will not be discussed in this book.

6.1.3 Local phase

Most stochastic methods of global optimization differ in the way in which starting points for local methods are chosen. This part of the local phase will be described separately for each optimization method that is discussed in this chapter.

Remarks to local methods that are used in the local phase

In theoretical considerations about the properties of local methods, a frequent assumption is that local procedures are - in the case of minimization - *strictly descent* (see Chapter 2, Definition 2.4 and also Rinnooy Kan, Timmer [137]).

In practice, it is difficult to check if an optimization procedure is strictly descent. Instead, ε-*descent* procedures can be considered.

Definition 6.1. *A local method* loc *will be called* ε*-descent on* $\mathcal{D}$ *if, for each starting point* $x_0 \in \mathcal{D}$ *and an arbitrary norm* $\|\cdot\|$ *in* V *it generates a sequence* $\{x_i\}, i = 1, 2 \ldots \subset \mathcal{D}$ *so that*

$$x_{i+1} = x_i + \alpha_i p_i \ \forall i = 0, 1, 2 \ldots, \ \ p_i \in V, \ \|p_i\| = 1, \ \alpha_i \geq 0 \,. \tag{6.3}$$

Moreover, the sequence converges to $x^+ \in \mathcal{D}$ *being a local minimizer, and satisfies*

$$\begin{aligned} &\forall i = 0, 1, 2 \ldots \ \forall j = 1, 2, \ldots, \forall \varepsilon > 0 \\ &\mathrm{Int}\left(\frac{\alpha_i}{\varepsilon}\right) \tilde{\Phi}(x_i + j\,\varepsilon\, p_i) \leq \tilde{\Phi}\left(x_i + (j-1)\,\varepsilon\, p_i\right) \,. \end{aligned} \tag{6.4}$$

□

Let us introduce new terms (see Rinnooy Kan, Timmer [137]). For x^+ being a local minimizer of the function $\tilde{\Phi}$ on $\mathcal{D}$

- $\mathcal{B}^{\varepsilon}_{x+} = \{x \in \mathcal{B}_{x^+} : \exists x_1 \in \mathcal{D} \setminus \mathcal{B}_{x^+} \text{ and } d(x, x_1) \leq \varepsilon\}$,
- $y_\varepsilon = \inf_{x \in \mathcal{B}^{\varepsilon}_{x+}} \tilde{\Phi}(x)$ if $\mathcal{B}^{\varepsilon}_{x+}$ is nonempty, $y_\varepsilon = \max(\tilde{\Phi}(x))$ otherwise,
- $\tilde{\mathcal{B}}^{\varepsilon}_{x+} = \left\{x \in \mathcal{B}_{x^+} : \tilde{\Phi}(x) < y_\varepsilon\right\}$.

Remark 6.2. (see Rinnooy Kan, Timmer [137]) Every ε-descent method loc has the following features:

1. Let $x \in \mathcal{D}$ be an arbitrary admissible point and $y \leq \tilde{\Phi}(x)$. If there is no point $x_1 \in L(y)$, $x_1 \notin L_x(y)$ such that the distance from x_1 to an element of $L_x(y)$ is less or equal to ε, then $\mathrm{loc}(x_0) \in L_x(y)$ for all starting points $x_0 \in L_x(y)$.
2. $\mathrm{loc}(x_0) = x^+$ for all starting points $x_0 \in \tilde{\mathcal{B}}^{\varepsilon}_{x+}$. □

The assumption that procedures are ε-descent simplifies theoretical considerations but does not influence significantly the main results of them. In practice, it may be that local methods are not strictly descent and are not even ε-descent. Moreover, even ε-descent methods can be not convergent for some objective functions and large values of ε. In the case of non-convergence one can try to change the parameters of the local method or choose another method.

6.1.4 Pure Random Search (PRS), Single-Start, Multistart

Pure Random Search (PRS) (see Chapter 2, Section 2.2) is a simple stochastic method used to find global minimizers or maximizers. It does not contain the local phase and it does not contain any mechanism which could be used to find local minima. PRS can be used for solving problems $\mathbf{\Pi_1}$ or $\tilde{\mathbf{\Pi}}_\mathbf{1}$. In the version that is presented below, each population contains only one point.

1: $t \leftarrow 1$; $y_0 \leftarrow \infty$
2: **repeat**
3: generate $x \in \mathcal{D}$ (use the uniform distribution)
4: **if** $\tilde{\Phi}(x) < y_{t-1}$ **then**
5: $y_t \leftarrow \tilde{\Phi}(x)$; $x_t \leftarrow x$
6: **else**
7: $y_t \leftarrow y_{t-1}$; $x_t \leftarrow x_{t-1}$
8: **end if**
9: $t \leftarrow t+1$
10: **until** the stopping rule is satisfied

Algorithm 2: Pure Random Search

PRS algorithm (version for minimization problems)

A modification of the above algorithm that involves a local method which starts from the best point (after a certain number of generations) is called Single-Start.

Another modification that consists of starting local methods from each sample point is called Multistart . Multistart detects minimizers, so it can be used to solve problems $\{\tilde{\mathbf{\Pi}}_i^{\mathbf{a}j}\}, i = 1, 2, 3; j = 1, 2, 3$. This method gave rise to clustering methods that are presented in Section 6.1.6.

6.1.5 Properties of PRS, Single-Start and Multistart

Let us define the new probability space $(\Omega^\infty, \Sigma_\Omega^\infty, P^\infty)$, where $\Omega^\infty = \Omega \times \Omega \times \ldots$, by Σ_Ω^∞ we denote the smallest σ-algebra that contains the algebra of cylindric sets. Here the elementary event is an infinite sequence of drawings of points from $\mathcal{D}$. The existence and construction of P^∞ follows from the Kolmogorov theorem of conformation (see [28]).

Let's define the following random variables for $k = 1, \ldots$:

$$\begin{aligned} &\underline{y}_k : \Omega^\infty \to \mathbb{R} \\ &\underline{y}_k((\omega_1, \omega_2, \ldots, \omega_k)) = \min\left\{\tilde{\Phi}(\underline{x}_1(\omega_1)), \tilde{\Phi}(\underline{x}_2(\omega_2)), \ldots, \tilde{\Phi}(\underline{x}_k(\omega_k))\right\}, \end{aligned} \tag{6.5}$$

$\omega_i \in \Omega$, $i = 1, 2, \ldots, k$. Random vectors $\underline{x}_i$ have been defined in section 6.1.1.

PRS has the following property (see [59], [137], see also Proposition 1 in [83]):

Theorem 6.3. *For* $\tilde{\Phi} \in C(\mathcal{D})$

$$P^\infty\left\{\lim_{k\to\infty} \underline{y}_k = \tilde{\Phi}^*\right\} = 1 \tag{6.6}$$

where $\tilde{\Phi}^*$ *denotes the global minimum of* $\tilde{\Phi}$ *on* $\mathcal{D}$.

Similar property has been proved for Single-Start and Multistart. For Multistart the random variables y_k have to be redefined in the following way:

$$\underline{y}_k : \Omega^\infty \to \mathbb{R}$$

$$\underline{y}_k\left((\omega_1, \omega_2, \dots, \omega_k)\right) = \min\left\{\text{loc}(\underline{x}_1(\omega_1)), \text{loc}(\underline{x}_2(\omega_2)), \dots, \text{loc}(\underline{x}_k(\omega_k))\right\}. \tag{6.7}$$

PRS, Single-Start and Multistart find the global extremum with the probability equal to 1 if the size of the sample goes to infinity. These methods are asymptotically correct in the probabilistic sense, according to Definition 2.16. For PRS and Multistart there are known optimal and sub-optimal stop criterions (see [59, 208, 32, 30, 23]). These criterions will be discussed in Section 6.2.

Multistart is an interesting method because of its mentioned good features, however, it is slow. The obvious drawback of Multistart is that it finds the same local extremum many times. One class of methods which aim at diminishing this drawback and accelerating computations are clustering methods. These methods are more efficient and they are also asymptotically correct.

6.1.6 Clustering methods in continuous global optimization

General description

Generally, *clustering methods* (see e.g. [89] and references therein) constitute a specific group of pattern recognition techniques, which do not utilize any learning set. Clustering consists of unsupervised exploration of a given data set, aimed at discovering groups in the data. To be more formal, clustering results in the construction of a partition of a discrete data set $X = \{x_1, \dots x_m\}$ into non-empty, exclusive subsets $X_1, \dots X_k$; $k \leq m$ called *clusters* . That means clustering is governed by some equivalence relation R on X and all data items classified into the same cluster X_i are equivalent in the sense of R. Clustering algorithms has also been applied in continuous global optimization. A group of such methods is described in Guus, Boender, Romeijn [83]. The idea of applying clustering methods in global optimization is to determine groups of points from which local searches can be started.

By clusters in global optimization we can understand sets of points that belong to approximations of sets of attraction of local minimizers. Clustering methods can be used in order to detect all local minima or the global minimum (together with minimizers). What is important, these methods enable us also to approximate the attraction sets of local minimizers. The idea of clustering methods described in [137] and [138] is to employ a local method only once in the basin of attraction of each detected local minimizer (under the assumption that local methods are strictly-descent or ε-descent, see Section 6.1.6). A similar idea can be seen in Multi Level Single Linkage (MLSL) which has been derived from clustering methods. However, in this method clusters are not detected. Another path in the research of clustering methods in global optimization has been developed by Adamska (see [152]). In her approach, a clustering method known from data analysis (FMM) is applied to continuous global optimization. This method gives the estimation of the density of measures. Clusters correspond to level sets of the measure density and can be represented by ellipsoids. This gives the possibility to approximate central parts of the basins of attraction of local extrema.

The common schema for clustering methods described in this chapter is as follows:

- generate m random points from the admissible set $\mathcal{D}$ according to the uniform distribution,
- transform the set of random points in order to facilitate cluster recognition,
- determine (recognize) clusters,
- in each cluster start a local method and store the result.

This schema can be repeated until a global stopping rule is satisfied.

The good point of clustering methods in global optimization is that the number of local searches (which are time consuming) is diminished. This is especially significant when the function evaluation is expensive.

Transformations of the set of random points

Guus, Boender and Romeijn in the monograph [86] mention two ways in which random points in global optimization clustering methods can be transformed: *reduction* and *concentration*. Reduction consists of the rejection of those random points (generated previously) for which the objective value is greater than a certain threshold (see Section 2.2 and also Becker, Lago [20]). The set of remaining points, called a reduced sample, naturally approximate a level set of the objective function. Often this level set is not simply connected. Concentration consists of starting several steps of a simple local method (for instance the gradient method, see Törn [189], the method of random directions, see Wit [202]). Telega proposed another concentration method, through the use of genetic algorithms (see Telega [184], Cabib, Schaefer, Telega [43]). This method called Clustered Genetic Search (CGS) is described in Section 6.3.

Cluster determination

Cluster determination begins with a *seed point*. New points are attached to the cluster according to *clustering rules*. The seed can be this unclustered point (i.e. a point which does not belong to any cluster yet) which has the smallest objective value or a point obtained as the result of a local method (often simplified) that starts from the sample point with the smallest objective value. Popular clustering rules and clustering methods are Density Clustering (DC) (see Guus, Boender, Romeijn [83], see also Törn [189], Wit [202]), Single Linkage (SL) (see Rinnooy Kan, Timmer [137], see also Guus, Boender, Romeijn [83]). Multi Level Single Linkage (MLSL) is not considered to be a clustering rule or a clustering method, however, it is derived from SL.

Analysis of clustering methods and clustering rules given by Rinnooy Kan and Timmer [137, 138] has been based on the assumption that the uniform distribution is used and the sample is transformed by the reduction. This approach enables us to use Bayesian methods in deriving stopping rules, analogously to Multistart (see Section 6.2). The next sections contain a description of the reduction phase and a description of scalar versions of DC, SL and MLSL together with a short analysis of them.

6.1.7 Analysis of the reduction phase

Terms

We assume the same terms as in sections 6.1.1 and 6.1.3. Additionally, let $y_k^{(i)}$ denote the ith value of the objective function from a sample of the size $k\,m$ (k drawings, m points in each drawing, no reduction). Values of the objective are ordered ascendingly.

One can expect that clustering methods should be quicker than Multistart. It would be desirable for clustering methods to have similar good properties like the asymptotic guarantee of success or the asymptotic correctness. Additionally, stopping rules should be theoretically justified. Such rules based on Bayesian analysis have been elaborated for Multistart. They will be described in Section 6.2. Analogous stopping rules have been proposed for clustering methods in global optimization. Such clustering methods can be treated as methods derived from Multistart, in which the sample is transformed, for instance by the reduction, and local methods are only started from some (unlike in Multistart) points of the transformed sample. Considerations about stopping rules for clustering methods will be preceded by the analysis of modifications that have been introduced in clustering methods in comparison to Multistart. This section contains a short analysis of the reduction phase.

Most Bayesian stopping rules for Multistart (non-sequential, sequential optimal and suboptimal) are based on the knowledge about the size of the sample and the number of local minimizers detected. Such stopping rules can be applied to every method, which if it starts from the same set of points as Multistart gives the same set of local minimizers (gives the same results). In particular, they can be applied for methods in which a local method starts exactly once in the attraction sets of local minimizers that contain points from the sample.

The Bayesian analysis (see Section 6.2) could be applied almost without changes if the set $L(y_\gamma)$ is considered instead of the admissible set $\mathcal{D}$. For any

$0 < \gamma < 1$, y_γ is calculated from the equation:

$$\frac{\text{meas}\left(\left\{x \in \mathcal{D} : \tilde{\Phi}(x) \leq y_\gamma\right\}\right)}{\text{meas}(\mathcal{D})} = \gamma. \tag{6.8}$$

However, in practice it is difficult or even impossible to determine the set $L(y_\gamma)$. Assume that after the reduction, $\gamma\, k\, m$ points are left (actually $\lfloor \gamma k\, m \rfloor$ or $\lceil \gamma\, k\, m \rceil$, however, by omitting these symbols the analysis can be simplified and the main results are the same). So the set which is considered can be defined as $L(y_k^{(\gamma\, k\, m)})$. This set changes with k, so a simple application of the Bayesian analysis (where the admissible set does not change) is not possible. However, Rinnooy Kan and Timmer show after Bahadur that the random variable $\underline{y}_k^{(\gamma\, k\, m)}$ (defined similarly as in Section 6.1.5) converges with probability to y_γ while k increases, so Bayesian stopping rules can be applied as if $L(y_k^{(\gamma\, k\, m)})$ does not change with k, at least for a large number of iterations (see Rinnooy Kan, Timmer, [137], Bahadur [18]).

The problem of how to determine points from which local methods can start will be described in subsequent sections together with clustering rules. Ideally, a local method should be started exactly once in each attraction set of a local minimizer, in which there is at least one point from the reduced sample. Then the method would give the same set of local minimizers as Multistart (on the set $L(y_k^{(\gamma\, k\, m)})$). This would allow us to apply the same stopping rules as in Multistart. Of course the total cost of local searches for such a method would be significantly reduced in comparison to Multistart.

Remark 6.4. Application of Multistart to the reduced sample (to the sets $L(y_k^{(\gamma\, k\, m)})$) implies the asymptotic probabilistic guarantee of success (Definition 2.16) , so the reduction phase does not affect this property of clustering methods. □

Remark 6.5. However, none of the clustering methods that contain the reduction phase has the property of asymptotic probabilistic guarantee of success in the sense of finding all local minimizers (Definition 2.17), regardless of the way in which starting points for local methods (in the local phase) are determined. The local phase may even worsen the effectiveness by omitting some local minimizers that could be found by Multistart with the reduction phase. It is a natural consequence of the rejection of this part of the domain of searches in which values of the objective function are greater than a constant (approximately y_γ). □

It is possible that a cluster recognized with the use of the clustering method in fact contains more than one minimizer.

The DC and SL clustering rules presented below are constructed in such a way that the probability that a local method will be not applied to a point that would lead to an undiscovered local minimizer diminishes to zero when the size of the sample grows.

6.1.8 Density Clustering

Let k be the number of drawings of m points, loc denotes the local method, x_s is the seed of a cluster, $C(x_s)$ denotes the cluster with the seed x_s, X^+ is the set of local minimizers found, $\#X^+$ stands for the number of elements in X^+. The following algorithm requires the determination of the parameters r_i, $i = 1, 2, ...$ according to Formula 6.17, which will be defined later in this section.

Algorithm

Construction of T_i sets, asymptotic properties of DC

In this method, clusters are built in a stepwise manner. First, a seed x_s is determined. It can be the "best" unclustered point $\overline{x}$ from the reduced sample, like in Törn [189] and Wit [202] or a point that is obtained from a local method loc started from $\overline{x}$. Let us denote by T_0 the set that contains only the seed of a cluster. In subsequent steps denoted by i, new points from the reduced sample are attached to the cluster. These points belong to the sets T_i, $i = 1, 2, \ldots$ which are subsets of the admissible set $\mathcal{D}$, $T_i \subset T_{i+1}$, $i = 1, 2, \ldots$. In simple versions of DC the sets T_i, $i = 1, 2, \ldots$ are balls. The radius of the ball is increased in subsequent steps until the density of points from the reduced sample is greater than a certain constant, which is, for instance, equal to the average density of the sample points in the admissible set $\mathcal{D}$ without the reduction. Such a version has been proposed by Törn [189]. In the version proposed by Rinnooy Kan and Timmer (see [137]) the shape and the volume of T_i, $i = 1, 2, ...$ can be determined after the following reasoning. A cluster which is initiated by the seed x_s should be related to the set $L_{x_s}(y_k^{(\gamma\, k\, m)})$. This suggests that subsequent T_i should be approximations of $L_{x_s}(y)$ for y increased stepwise. Under the assumption that $\tilde{\Phi} \in C^2(\mathcal{D})$ we can approximate the level sets using the estimation $\tilde{\Phi}(x) \approx \tilde{\Phi}(x_s) + \frac{1}{2}(x - x_s)^T H(x_s)(x - x_s)$, where H denotes Hessian. Hence T_i can be defined as follows:

$$T_i = \{x \in \mathcal{D} : (x - x_s)^T H(x_s)(x - x_s) \le r_i^2\} \tag{6.9}$$

for $r_i < r_{i+1}$, $i = 1, 2, \ldots$.

In this approach T_i are ellipsoids. An approximation of Hessian can be easily obtained as a by-product of quasi-Newton local methods. Radiuses r_i are set in such a way that guarantees the following asymptotic property: when k increases, the probability that the process of cluster recognition is stopped too early is less or equal to the probability that there is no original sample (i.e. not reduced sample) point in $\Delta T_i = \{x \in \mathcal{D} : x \in T_i, x \notin T_{i-1}\}$. One can say that the process is stopped too early if, in the set ΔT_i, there is no point of the reduced sample but there is at least one such point in $L_{x_s}(y_k^{(\gamma\, k\, m)})$. This situation has been called an error of type I by Boender (see [34]).

```
1: k ← 0; X+ ← ∅;
2: repeat
3:    k ← k + 1
4:    /* Determination of reduced sample */
5:    Draw m points x_{(k-1)m+1}, ..., x_{km} according to the uniform distribution on
      D. Choose γkm "best" points from the cumulated sample x_1, ..., x_{km};
6:    j ← 1; /* Number of the cluster that is being recognized */
7:    while Not all points from the reduced sample have been assigned to clusters;
      do
8:       /* Determination of the seed */
9:       if j ≤ #X+ then
10:         Choose j^th local minimizer in X+ as the seed x_s of the cluster C(x_s)
11:      else
12:         repeat
13:            x̄ ← unclustered point from the reduced sample with the smallest
               objective value;
14:            Apply a local method loc to x̄;
15:            x+ ← loc(x̄);
16:            if x+ ∈ X+ then
17:               C(x+) ← C(x+) ∪ {x̄};
18:            end if
19:         until x+ ∉ X+;
20:         X+ ← X+ ∪ {x+};
21:         x_s ← x+;
22:      end if
23:      i ← 0; T_i ← {x_s};
24:      repeat
25:         /* New points are added to the cluster that is being recognized */
26:         i ← i + 1;
27:         Add to C(x_s) all unclustered points that are located in the set T_i with
            parameter r_i(x_s) (see formulas 6.9 and 6.17);
28:      until No new point has been added to C(x_s);
29:      j ← j + 1;
30:   end while
31: until The global stopping rule is satisfied;
```

Algorithm 3: Density Clustering

The probability of an error of type I in step i is less than or equal to α_k, where

$$\alpha_k = \left(1 - \frac{\text{meas}\,(\Delta T_i)}{\text{meas}\,(\mathcal{D})}\right)^{k\,m}. \tag{6.10}$$

The set ΔT should contain at least one point of the not reduced sample with the probability $(1 - \alpha_k)$, under the assumption that the distribution is uniform.

From Formula 6.10 we have

$$\text{meas}(\Delta T_i) = \text{meas}(\mathcal{D})\left(1 - \alpha_k^{\frac{1}{k\,m}}\right) \tag{6.11}$$

so when we assume that the probability of an error of type I is less than or equal to α_k then we can expect that a set whose volume is given by Formula 6.11 contains at least one sample point. We assume that T_i are ellipsoids. The volume of the ellipsoid $(x - x_s)^T H(x_s)(x - x_s) \leq r_i^2$ is equal to

$$\frac{\pi^{\frac{N}{2}} r_i^N}{\Gamma\left(1 + \frac{1}{2}N\right) \det\left(H(x_s)\right)^{\frac{1}{2}}} \tag{6.12}$$

where Γ stands for Gamma Euler's function (see for instance [28] Chapter 18). Hence, the stopping rule can be as follows: no unclustered point x from the reduced sample has been found, for which

$$\begin{gathered}\pi(x - x_s)^T H(x_s)(x - x_s) \leq \\ \left[i\Gamma\left(1 + \tfrac{1}{2}N\right) \det\left(H\left(x_s\right)\right)^{\frac{1}{2}} \operatorname{meas}(\mathcal{D}) \left(1 - \alpha_k^{\frac{1}{k\,m}}\right)\right]^{\frac{2}{N}}.\end{gathered} \tag{6.13}$$

Rinnooy Kan and Timmer have proposed in [137] such a ΔT_i that

$$\operatorname{meas}(\Delta T_i) = \operatorname{meas}(\mathcal{D}) \frac{\sigma \log(k\,m)}{k\,m}, \ \sigma > 0. \tag{6.14}$$

Hence

$$\alpha_k = \left(1 - \frac{\sigma \log(k\,m)}{k\,m}\right)^{k\,m}. \tag{6.15}$$

The probability α_k decreases polynomially with increasing k, because $\forall k$ $\exists c_1, c_2$ such that

$$c_1 k^{-\sigma} \leq \left(1 - \frac{\sigma \log k}{k}\right)^k \leq c_2 k^{-\sigma}. \tag{6.16}$$

Thus, the parameter r_i can be evaluated from the formula:

$$r_i = \pi^{-\frac{1}{2}} \left(i\Gamma\left(1 + \frac{N}{2}\right) \det\left(H\left(x_s\right)\right)^{\frac{1}{2}} \operatorname{meas}(\mathcal{D}) \frac{\sigma \log(k\,m)}{k\,m}\right)^{\frac{1}{N}}. \tag{6.17}$$

Remark 6.6.

1. In DC if the process of cluster recognition is stopped in step i, i.e. there is no point in T_i (the parameter r_i is calculated from Formula 6.17), then the probability that a cluster has been recognized improperly with an error of type I decreases polynomially while k increases. This follows directly from the discussion presented above.
2. The basic disadvantage of this version of DC is that the objective function is approximated by the square function and the attraction sets are approximated by ellipsoids. If the set $L_{x^+}\left(y_k^{(\gamma k m)}\right)$ differs significantly from an ellipsoid, then the criterion concerning when to stop the cluster

recognition process is incorrect (other estimations given in this section are also incorrect). A good clustering method should more accurately approximate the sets $L_{x^+}\left(y_k^{(\gamma\, k\, m)}\right)$ with increasing k. One such method is Single Linkage. It is presented in the next section.

3. Similarly to PRS and Multistart, DC also has the property of asymptotic correctness (i.e. it finds the global minimum with the probability 1 with k increasing to infinity). This follows from the fact that the global phase has properties of PRS applied to the sets $L\left(y_k^{(\gamma\, k\, m)}\right)$.
4. We cannot expect all local minimizers to be found. Points of recognized clusters in a sense approximate some subsets of $\mathcal{D}$, they should be central parts of sets of attraction of local minimizers. However, in reality, these subsets may contain more than one local minimum and more minimizers. A cluster can cover more than one minimizer. Because the local method is started exactly once in each cluster, some local minimizers may remain undiscovered. Also the reduction phase itself can be a reason why some local minimizers could be omitted. DC does not have the property of the asymptotic probabilistic guarantee of finding all local minimizers.
5. The version presented in this section after Rinnooy Kan and Timmer [137] has the property of the asymptotic probabilistic guarantee of finding one local minimizer in each simply connected component of the set $\left\{x \in \mathcal{D} : \tilde{\Phi}(x) \leq y_\gamma\right\}$, where y_γ is calculated according to Formula 6.8 (when we assume that the local method is ε-descent, this is true if the components are not too close). □

The last conclusion requires comments. Boender proposed use of an additional criterion when points are assigned to a cluster (see [29]). This can be applied in cases of differentiable objective functions. An approximated derivative of $\tilde{\Phi}$ in the direction x_s is equal to

$$\frac{\tilde{\Phi}\left(x + h(x_s - x)\right) - \tilde{\Phi}(x)}{h\|x_s - x\|_2} \tag{6.18}$$

($\|\cdot\|_2$ denotes Euclidean norm on V) for small h is calculated. If the value is positive, then x is not assigned to the cluster. It is easy to give examples which show that this criterion can fail and a point will not be assigned to the cluster, though the point belongs to the set of attraction of the local minimizer. For instance, in a “spiral valley” the value of the derivative of $\tilde{\Phi}$ in x in the direction x_s can be positive, though any descent local method (e.g. gradient methods) converges to x_s. It is possible that such an error could be fixed in the next steps and the unclustered point would be assigned to the appropriate cluster. However, the criterion can also fail because it is possible that a point is assigned to one cluster, but it should be assigned to another one (because it belongs to the set of attraction of another local minimizer).

Another comment to the terminology should also be given. Zieliński (see [208]) defined the basins of attraction for local minima, not local minimizers. However, he considered only cases of isolated minimizers. Our definition of the basin of attraction (see Definition 2.6) refers to local minimizers (not minima), which are isolated. For isolated minimizers, both definitions can be used interchangeably.

6.1.9 Single Linkage

This method differs from DC in the way in which basins of attraction of local minimizers are approximated. Let the distance from a point x to a set A be defined as:

$$\rho(x, A) = \inf_{y \in A} ||x - y||_2. \tag{6.19}$$

P_k will stand for the set of all points from the reduced cumulated sample and $C = \bigcup_{x_s} C(x_s)$ will denote the set of points that have already been assigned to clusters.

Algorithm

Analysis of the properties of Single Linkage can be found in [137]. It is much more complicated than the analysis of the properties of Density Clustering, so here only the final results will be recalled (see Theorem 8 and Theorem 12 in [137]).

Theorem 6.7. *If the critical distance r_k is given by the formula*

$$r_k = \pi^{-\frac{1}{2}} \left(\Gamma\left(1 + \frac{N}{2}\right) \mathrm{meas}(\mathcal{D}) \frac{\sigma \log(k\,m)}{k\,m} \right)^{\frac{1}{N}} \tag{6.20}$$

then:

- *for $\sigma > 2$ the probability that the local method will be started by SL in step k tends to zero with increasing k,*
- *for $\sigma > 4$ even if the sampling of points had been carried endlessly, the total number of local searches started by SL would be finite with the probability equal to 1.* □

Theorem 6.8. *If r_k tends to zero with increasing k, then in each component of $L(y_\gamma)$ that contains points from the sample, the local minimizer will be found with the probability equal to 1 in a finite number of steps.* □

Remark 6.9.

1. The disadvantage of DC related to the way in which the objective function and the basins of attraction are approximated is eliminated in SL.

1: $k \leftarrow 0$; $X^+ \leftarrow \emptyset$;
2: **repeat**
3: $k \leftarrow k+1$
4: /* Determination of reduced sample*/
5: Draw m points $x_{(k-1)m+1}, ..., x_{k\,m}$ according to the uniform distribution on $\mathcal{D}$. Choose $\gamma\, k\, m$ "best" points from the cumulated sample $x_1, ..., x_{k\,m}$;
6: $j \leftarrow 1$; /* Number of the cluster that is being recognized */
7: **while** Not all points from the reduced sample have been assigned to clusters; **do**
8: /* Determination of the seed */
9: **if** $j \leq \#X^+$ **then**
10: Choose j^{th} local minimizer in X^+ as the seed x_s of the cluster $C(x_s)$
11: **else**
12: **repeat**
13: $\overline{x} \leftarrow$ unclustered point from the reduced sample with the smallest objective value;
14: Apply a local method loc to $\overline{x}$;
15: $x^+ \leftarrow loc(\overline{x})$;
16: **if** $x^+ \in X^+$ **then**
17: $C(x^+) \leftarrow C(x^+) \cup \{\overline{x}\}$;
18: **end if**
19: **until** $x^+ \notin X^+$;
20: $X^+ \leftarrow X^+ \cup \{x^+\}$;
21: $x_s \leftarrow x^+$;
22: **end if**
23: $l \leftarrow 0$;
24: **repeat**
25: /* New points are added to the cluster that is being recognized */
26: **if** $P_k \backslash C \neq \emptyset$ /* There are unclustered points (P_k stands for the reduced sample in the step k)*/ **then**
27: Find $x_l^k \in P_k \backslash C$ such that $\rho\left(x_l^k, C(x_s)\right) = \min_{y \in P_k \backslash C} (y, C(x_s))$;
28: **if** $\rho\left(x_l^k, C(x_s)\right) \leq r_k$ **then**
29: $C(x_s) \leftarrow C(x_s) \cup \{x_l^k\}$ /* x_l^k is assigned to the cluster $C(x_s)$ */
30: **end if**
31: **end if**
32: $l \leftarrow l+1$
33: **until** $\rho\left(x_l^k, C(x_s)\right) > r_k$ OR $P_k \backslash C = \emptyset$;
34: $j \leftarrow j+1$;
35: **end while**
36: **until** The global stopping rule is satisfied;

Algorithm 4: Single Linkage

2. The number of local searches can be treated as an effectiveness measure of the method. Theorems 6.7 and 6.8 show that SL has good asymptotic properties. The number of local searches is finite even if the sampling of points is carried out endlessly.
3. Rinnooy Kan and Timmer in [137] mention experiments which show that SL better approximates sets $L_{x^*}\left(y_k^{(\gamma\, k\, m)}\right)$ than DC does.
4. SL has the property of asymptotic correctness, i.e. it finds the global minimum with the probability equal to 1 when k increases to infinity. It follows from Theorem 6.8 when we assume that r_k is set according to Formula 6.20
5. SL does not have the property of probabilistic guarantee of success in the sense of finding all local minimizers.
6. Multi Level Single Linkage, which has been derived from the clustering methods presented above, has the property of probabilistic guarantee of success - it can find all local minimizers. This follows from Theorem 6.8.

□

6.1.10 Mode Analysis

Mode Analysis (MA) has been derived from Single Linkage. The main difference between SL and MA is that greater subsets of the domain are assigned to clusters than single points. In the version which is described below these subsets are hypercubes (see Rinnooy Kan, Timmer [137]).

Terms and definitions

For the sake of simplicity let us assume that $\mathcal{D}$ is a hypercube that can be divided into κ hypercubes ($\sqrt[N]{\kappa}$ is an integer, $\mathcal{D} \subset \mathbb{R}^N$).

- Hypercubes that are results of the division of $\mathcal{D}$ will be called *cells*.
- A cell U will be called *full* if it contains more than

$$\frac{\mathrm{meas}(U) k\, m}{2\, \mathrm{meas}(\mathcal{D})} \tag{6.21}$$

  points from the reduced sample.
- A cell which is not full will be called *empty*.
- Two cells U_a and U_b will be called *neighboring* if $\forall \varepsilon > 0\ \exists x_a \in U_a$ and $x_b \in U_b$ so that $||x_a - x_b|| < \varepsilon$.
- Clusters will contain these points from the reduced sample that belong to neighboring full cells. The cluster recognition begins with a *seed-cell*, which contains a point from the reduced sample with the smallest objective value

(like in [137]). A seed-cell can also be the cell which contains the "best" minimizer found.

- By $C(U_s)$ we denote the cluster, for which the seed cell is U_s.
- In the original approach clusters are sets of points (after the reduction). However, one could also consider such a version of *MA*, in which clusters are unions of whole hypercubes, which are full cells. Depending on the version by *assigning* the cell U to a cluster C we can understand:
  1. assigning of these points from the reduced sample, that belong to U
  2. attaching the whole cell U to cluster C (union). □

Algorithm

Rinnooy Kan and Timmer compared Single Linkage and Mode Analysis (see [137]). The comparison shows that one cannot state which method is better. In some cases SL recognizes clusters incorrectly while MA recognizes them correctly and vice versa.

We recall here two theorems concerning asymptotic properties of the Mode Analysis method (see Theorem 14 and Theorem 15 in [137]).

Theorem 6.10. *If the number of cells in MA is equal to*

$$\kappa = \frac{k\,m}{\sigma \log(k\,m)} \tag{6.22}$$

then

- *for $\sigma > 10$ the probability that the local method will be started by MA in step k tends to zero with increasing k,*
- *for $\sigma > 20$ even if the sampling continues forever, the total number of local searches started is finite with the probability equal to 1.* □

Theorem 6.11. *If the number of cells in MA is equal to κ which is given by Formula 6.22 (for $\sigma > 0$) then for each component $L(y_\gamma)$ which contains points from the sample, the local minimizer will be found in a finite number of steps with the probability equal to 1.* □

Remark 6.12.
1. All remarks given for SL are also true for MA.
2. Because MA makes use of cells it is more immune to errors related to random irregularity of the sample (that means the local deviation from the uniform covering). However, on the other hand it is possible that in MA points from the sample will be incorrectly assigned to clusters because of too large cells.
3. MA enables us to relatively easily approximate the sets of attraction of local minimizers by a union of hypercubes. □

```
 1: k ← 0; X⁺ ← ∅
 2: repeat
 3:    k ← k + 1
 4:    /* Determination of the reduced sample */
 5:    Draw m points x_{(k-1)m+1}, ..., x_{k m} according to the uniform distribution on
       D. Choose γ k m "best" points from the cumulated sample x_1, ..., x_{k m};
 6:    Divide D into κ cells.
 7:    /* Determination of cells, determination of which cells are full and which are
       empty */
 8:    For each cell determine how many points from the reduced sample are inside
       the cell. If this number is greater than the value given by Formula 6.21 then
       the cell is full, otherwise the cell is empty.
 9:    j ← 1; /* Number of the cluster that is being recognized */
10:    while Not all full cells have been assigned to clusters; do
11:       /* The determination of the seed of a cluster. The seed is a full cell.*/
12:       if There is an unclustered full cell U_s which contains an element from X⁺
          (i.e. it contains a local minimizer) then
13:          Choose U_s as the seed of the cluster C(U_s);
14:          while There is an unclustered full cell U_b which is a neighbor of any cell
             assigned to cluster C(U_a) do
15:             Assign U to C(U) (assign points from U to C(U))
16:          end while
17:       else
18:          Determine the point x̄ so that the objective value in this point is the
             minimum value of the set of values in points which are in unclustered full
             cells.
19:          Apply the local method loc to x̄; x⁺ ← loc(x̄);
20:          X⁺ ← X⁺ ∪ {x⁺};
21:       end if
22:       j ← j + 1;
23:    end while
24: until The global stopping rule is satisfied;
```

Algorithm 5: Mode Analysis

6.1.11 Multi Level Single Linkage and Multi Level Mode Analysis

Multi Level Single Linkage (MLSL) has been derived from clustering methods (see [138, 83]). Unlike DC or SL it does not contain the reduction phase. Hence, there are no disadvantages caused by this phase.

Two versions of this method will be presented below. The first one contains a clustering process, the second one does not. The goal of both versions is to find all local minimizers. The second version cannot be applied when the sets of attraction of local minimizers are to be found. It has been proved that both versions have good asymptotic properties: the asymptotic probabilistic correctness and probabilistic guarantee of finding all local minimizers.

Algorithm version 1

1: $k \leftarrow 0$; $X^+ \leftarrow \emptyset$
2: **repeat**
3: $k \leftarrow k+1$
4: Draw m points $x_{(k-1)m+1}, ..., x_{k\,m}$ according to the uniform distribution on $\mathcal{D}$;
5: Sort all points drawn in k steps so that $\tilde{\Phi}(x_i) \leq \tilde{\Phi}(x_{i+1})$, $1 \leq i \leq k\,m-1$
6: $j \leftarrow 0$; /* The number of clusters that are already recognized */
7: $i \leftarrow 1$;
8: **while** $i \leq k\,m$ **do**
9: Add x_i to each cluster which contains points within the distance r_k from x_i;
10: **if** x_i is unclustered **then**
11: Start a local method loc from x_i;
12: $x^+ \leftarrow loc(x_i)$;
13: **if** $x^+ \in X^+$ **then**
14: $C(x^+) \leftarrow C(x^+) \cup \{x_i\}$
15: **else**
16: /* A new cluster has been found*/
17: $X^+ \leftarrow X^+ \cup \{x^+\}$;
18: $j \leftarrow j+1$;
19: Set the seed x_s of the new cluster $C(x_s)$ to x^+;
20: **end if**
21: **end if**
22: $i \leftarrow i+1$
23: **end while**
24: **until** The global stopping rule is satisfied;

Algorithm 6: Multi Level Single Linkage version 1

In this method one point from the sample can be assigned to more than one cluster. This reflects the fact that it is possible that there are such subsets of $\mathcal{D}$ that different local methods (even strictly descent) which start from these subsets can find different minimizers. Clusters initiated by x^+ are neither related to $R^{loc}_{x^+}$ nor to the basins $\mathcal{B}_{x^+}$. They are also not related to $L_{x^+}(y_k^{(\gamma\,k\,m)})$. Clusters are sets that contain such $x \in \mathcal{D}$, for which there is a r_k-descent sequence of points from the sample $x_1 = x, x_2, ..., x_\nu = x^+$ for a certain ν. A sequence is called r_k-descent if the distance from two subsequent elements is not greater than r_k and $\tilde{\Phi}(x_1) \geq ... \geq \tilde{\Phi}(x_\nu)$. If two different methods that start from the same point find different local minimizers there is a risk that some local minimizers remain undiscovered. However, it has been proved (see Rinnooy Kan, Timmer [138]) that this kind of error does not appear if local methods start from the interior of the basin $\mathcal{B}_{x^+}$. If k increases and r is suitably small this error does not appear with the probability 1 (see Theorem

6.13 and 6.14 below). The name of the method comes from the fact that it finds the same local minimizers as SL which is applied to the sets $L(y_k^{(i)})$ for each $k = 1, 2, ..., k\,m$, i.e. for subsequent cut-off levels which are determined by $y_k^{(i)}$.

Algorithm version 2

1: $k \leftarrow 0$; $X^+ \leftarrow \emptyset$
2: **repeat**
3: $k \leftarrow k + 1$
4: Draw m points $x_{(k-1)m+1}, ..., x_{k\,m}$ according to the uniform distribution on $\mathcal{D}$;
5: $i \leftarrow 1$;
6: **while** $i \leq k\,m$ **do**
7: **if** NOT (there is such a j that $\tilde{\Phi}(x_j) < \tilde{\Phi}(x_i)$ and $||x_j - x_i||_2 < r_k$) **then**
8: Start a local method loc from x_i;
9: $x^+ \leftarrow loc(x_i)$;
10: $X^+ \leftarrow X^+ \cup \{x^+\}$;
11: **end if**
12: $i \leftarrow i + 1$
13: **end while**
14: **until** The global stopping rule is satisfied;

Algorithm 7: Multi Level Single Linkage version 2

The asymptotic properties of both versions are the same. They are expressed by two theorems (see Theorem 1 and Theorem 2 in [138]):

Theorem 6.13. *If the critical distance r_k is given by the following formula:*

$$r_k = \pi^{-\frac{1}{2}} \left(\Gamma \left(1 + \frac{N}{2} \right) \text{meas}(\mathcal{D}) \frac{\sigma \log(k\,m)}{k\,m} \right)^{\frac{1}{N}} \tag{6.23}$$

then

- *if $\sigma > 0$, then for any sample point $x \in \mathcal{D}$ the probability that the local method will start from this point in step k decreases do zero while k increases,*
- *if $\sigma > 2$ then the probability that the local method will start in step k decreases to zero while k increases,*
- *for $\sigma > 4$ even if sampling is performed infinitely, the total number of local searches ever started by MLSL is finite with the probability 1.* □

Theorem 6.14. *If r_k tends to zero with increasing k, then each isolated local minimizer x^+ will be found by Multi Level Single Linkage with the probability 1 in a finite number of steps.* □

Remark 6.15.

1. Similarly to SL and DC, MLSL also has the property of asymptotic correctness (it finds the global minimum with probability 1 with k increasing to infinity). This follows from Theorem 6.14.
2. Similarly to SL, MLSL is effective in the sense that the total number of local searches started is finite with the probability equal to 1 (see Theorem 6.13).
3. Unlike DC and SL, MLSL has the property of probabilistic guarantee of success (in the sense of finding all local minimizers).
4. Usually the second version of MLSL is used, information about clusters is not gathered.
5. MLSL can be ineffective when the objective function has a large number of local minimizers which are close to each other and have small basins of attraction. MLSL will tend to start local methods in each basin, which is expensive. Local methods may wander around in shallow basins or areas similar to a plateau. This can be disadvantageous when only "essential" minimizers are to be found.
6. The effectiveness of MLSL depends on the appropriate value of the critical distance. This distance changes during computations. It may take a long time to achieve such values which guarantee that all local minimizers are discovered. □

Rinnooy Kan and Timmer also proposed another version of MLSL, in which hypercube cells are considered instead of single points (like in MA, see [137]). Properties of this method are similar to properties of the basic version of MLSL. Remarks analogous to items 2 and 3 from Remark 6.15 are true. This version enables us to approximate basins of attraction in a simple way.

6.1.12 Topographic methods (TGO, TMSL)

Topographical methods will be briefly described here for two reasons. First, their usefulness and superiority over MLSL for some global optimization problems has been proved (see Ali, Storey [1], Törn, Viitanen [190]). Second, it seems these methods can be modified in such a way that they can give information about the basins of attraction. Detailed modifications will not be proposed here. Only some capabilities will be indicated.

The idea of the Topographical Global Optimization (TGO) method is to utilize information contained in the so-called *topograph* . In the global phase, random points are generated from the uniform distribution. These sample points that are located too closely to points already accepted are rejected. In order to diminish time complexity, the so-called pre-sampling has been proposed. It consists of storing the accepted points of the sample (for instance on a disk), then this stored sample can be used with possible modifications (for instance, scaling to a domain of searches). The topograph is a directed graph, whose nodes represent sample points and edges connect the g nearest neighbors. The arrows are directed to nodes with a greater objective value. The *graph minima* are those nodes which do not have neighbors with a lower objective value. These minima should approximate minimizers of $\tilde{\Phi}$ at least if g is suitable. The problem of how to set the value of g has not been solved in a satisfactory way (see Törn, Viitanen [190]). The authors emphasize the high complexity of the method, in particular for multidimensional problems.

It seems that joined graphs of the g-nearest neighbors could be a source of information about the basins of attraction of local minimizers.

The second topographical method, Topographical Multilevel Single Linkage (TMSL) has been described by Ali and Storey in [1]. The authors propose the use of MLSL only for the minima of the topograph. This should accelerate the method significantly. The topograph here replaces the reduction phase (however, the reduction phase could also be employed in MLSL in order to accelerate computations). In the original version, clusters are not recognized and the proper determination of g is not as important as in TGO. When basins are to be found, the determination of g is important. For instance if g is equal to the sample size, only the global minimum can be found.

6.2 Stopping Rules

There are many papers and studies concerning the stopping rules of stochastic methods in global optimization. In this chapter the approach initiated by Zieliński (see [208]) will be presented. This approach has also been developed by Boender [29, 35], Rinnooy Kan [31, 32, 30], Betro, Shoen [23, 24] and others.

The following simple stopping rule could be employed in all stochastic methods in which random points are generated according to the uniform distribution on the admissible set $\mathcal{D}$. We assume that a point from A_ε is to be found (see Chapter 2, Section 2.1). The probability that a point from this set is generated during m drawings is equal to $1-(1-\mu(A_\varepsilon))^m$, where $\mu(A)$ for any measurable A stands for the relative Lebesgue measure

$$\mu(A) = \frac{\text{meas}(A)}{\text{meas}(\mathcal{D})}. \tag{6.24}$$

We can stop the algorithm if this probability is greater than $1-\delta$ for certain $\delta > 0$. Thus, the stopping rule can utilize the number of drawings m. This number should comply with the following inequality:

$$m \geq \frac{\log \delta}{\log(1-\mu(A_\varepsilon))}. \tag{6.25}$$

Such an approach has drawbacks. The set A_ε is not known (however, the measure $\mu(A_\varepsilon)$ could be approximated or could be set arbitrarily). Moreover, no additional information about the problem which is being solved is utilized. Also, the local phase is not considered.

Boender, Rinnooy Kan and Vercellis (see [33]) propose the following desirable properties of stopping rules:

- *Sample dependency*: either sample points or objective values in these points and also information about each local minimizer that is found (for instance how many times it was found) should be taken into consideration
- *Problem dependency*: information about the class of the objective function, the number of local minimizers (for instance, that it is less than a certain constant), the relative volumes of the basins of attraction of local minimizers should be taken into consideration.
- *Method dependency*: specific features of the algorithm should be taken into consideration.
- *Loss dependency*: costs of stopping too early (before the global minimum is found or before all "essential" minimizers are found) should be taken into consideration.
- *Resource dependency*: computational costs (time or memory) should be as small as possible. □

In this section the Bayesian approach to the stopping problem will be described. The probabilistic model is estimated from information gathered during the run of the optimization procedure. This model can be used to construct the so-called *non-sequential* stopping rules. Information about the costs of further searches is not used in this approach. However, such information is used in the so-called *sequential* rules which will also be described in this chapter.

The stopping rules presented below have been developed for Multistart. They can also be applied to some other clustering methods in global optimization. This chapter can be treated only as a basic survey of stopping rules in stochastic global optimization.

Because local methods are used, the probability of finding a local minimizer $x^+ \in \mathcal{D}$ is equal to the relative volume of its attraction set. In the Bayesian approach the number of local minimizers found is estimated together with relative volumes of their sets of attractions $\Theta_1, \Theta_2, ..., \Theta_w$. Let $\underline{w}$ be a random variable that is equal to the number of different minimizers. Let $\underline{\Theta}_1, \underline{\Theta}_2, ..., \underline{\Theta}_w$ be random variables that determine the relative measures of sets of attraction of local minimizers (see [32]). For the above random variables *a priori* distributions are assumed, then (based on the results of Multistart for m points, m stands for the sample size) *a posteriori* distributions are determined according to the Bayes rule. We assume that for the value of $\underline{w}$ every positive integer is equally probable, moreover, $\underline{\Theta}_1, \underline{\Theta}_2, ..., \underline{\Theta}_w$ have the uniform distribution on the $(w-1)$-dimensional unit simplex.

6.2.1 Non-sequential rules

Under the above assumptions the expected value (*a posteriori*) of a number of undiscovered minimizers is given by (see Boender [29], Guus, Boender, Romeijn [83]):

$$\mathrm{E}(\underline{w} - w) = \frac{w(w+1)}{m - w - 2} \tag{6.26}$$

(m stands for the number of points drawn, we assume that $m > w + 2$). The expected value (*a posteriori*) of the sum of relative volumes of the attraction sets of undiscovered minimizers is equal to:

$$\mathrm{E}\left(1 - \sum_{l=1}^{w} \underline{\Theta}_l\right) = \frac{w(w+1)}{m(m-1)}. \tag{6.27}$$

The probability (*a posteriori*) that all local minimizers have been found (i.e. $\underline{w} = w$) is equal to:

$$\Pr\{\underline{w} = w\} = \prod_{l=1}^{w} \left(\frac{m-1-l}{m-1+l}\right). \tag{6.28}$$

The following stopping rules can be derived from the above formulas.

- Stop if:

$$\mathrm{Int}\left(\frac{w(w+1)}{m-w-2}\right) = 0. \tag{6.29}$$

- Stop if the expected value of the sum of relative volumes of the attraction sets of undiscovered local minimizers is less than an arbitrary constant η_1:

$$\frac{w(w+1)}{m(m-1)} < \eta_1. \tag{6.30}$$

  This kind of stopping rule can be especially interesting when the objective function has many local minimizers with very small regions of attraction. Continuing the search in order to find *all* local minimizers can be extremely expensive (long) in such cases.
- Stop if the probability of finding all local minimizers is greater than an arbitrary constant η_2:

$$\prod_{l=1}^{w} \left(\frac{m-1-l}{m-1+l}\right) > \eta_2. \tag{6.31}$$

6.2.2 Sequential rules - optimal and suboptimal Bayesian stopping rules

Sequential stopping rules take into consideration the cost of sampling. Let us consider the two following *loss functions* (see Guus, Boender, Romeijn [83]):

$$L_1 = c_1^t(\underline{w} - w) + c_1^e m \tag{6.32}$$

and

$$L_2 = c_2^t\left(1 - \sum_{l=1}^{w} \Theta_l\right) + c_2^e m\,. \tag{6.33}$$

Coefficients $c_1^t, c_2^t > 0$ are related to the cost of stopping the algorithm before all local minimizers are found. This cost is called a *termination loss.* Coefficients $c_1^e, c_2^e > 0$ are related to the cost of continuation of the search. It is called an *execution loss.* Boender and Rinnooy Kan considered the above functions for $c_1^e = 1$ and $c_2^e = 1$, they gave also other loss functions (see [32]). The following results are given after Guus, Boender, Romeijn [83] and Boender, Rinnooy Kan [32].

Under the assumption that m points give w different minimizers, the expected values *a posteriori* of L_1 and L_2 (the so-called *expected posterior loss* or just the *posterior loss*) are equal to:

$$\mathrm{E}\left(L_1|(m,w)\right) = c_1^t\frac{w(w+1)}{m-w-2} + c_1^e m \tag{6.34}$$

$$\mathrm{E}\left(L_2|(m,w)\right) = c_2^t\frac{w(w+1)}{m(m-1)} + c_1^e m\,. \tag{6.35}$$

The pair (m, w) denotes that after m steps of the algorithm (m drawings and m local searches in Multistart) w different minimizers have been found.

The posterior loss after $m' > m$ observations is a random variable $E(L_i|(m', \underline{w})))$ (see [32]). A stopping rule is *optimal* if it minimizes the sum of $\mathrm{E}(L_i|(m', \underline{w}))$ for $m' = m+1, \ldots\infty$. For a pair (m, w) only two results of performing one more step are possible - either a new minimizer will be found (it is denoted by $(m+1, w+1)$) or the local method will converge to a known minimizer (this is denoted by $(m+1, w)$). The probability that the next step will finish without a new minimizer is equal to the expected value (*a posteriori*) of the sum of relative volumes of sets of attractions of undiscovered local minimizers (see Formula 6.30). The conditional expected value of the posterior loss of one more step of the algorithm can be estimated on the basis of 6.34 and 6.35 according to the recurrent formula ($i = 1, 2$):

$$\begin{aligned}\mathrm{E}\left(\mathrm{E}(L_i|(m+1), \underline{w}))|(m,w)\right) = {} & \left(1 - \frac{w(w+1)}{m(m-1)}\right)\mathrm{E}(L_i|(m+1,w)) \\ & + \frac{w(w+1)}{m(m-1)}\,\mathrm{E}(L_i|(m+1,w+1)).\end{aligned} \tag{6.36}$$

One more step of the algorithm means that one more point is generated and all other operations related to this point are performed, in Multistart it is the run of the local search.

The optimal rule stops the algorithm if the following criterion is satisfied

$$\mathrm{E}\left(\mathrm{E}(L_i|(m+1,\underline{w}))|(m,w)\right) > \mathrm{E}(L_i|(m,w)). \tag{6.37}$$

under the assumption that the optimal strategy would also be applied in subsequent steps. The estimation of the optimal rule begins with finding such m^* (if it exists), that for all pairs (m', w) where $m' \geq m^*$ Formula 6.37 holds (we assume here that $m \geq w+2$). In other words, for all (m', w), $m' \geq m^*$ the value of the expected posterior loss never decreases after the execution of the subsequent step. For all pairs (m', w) where $m' = m^*$ the optimal decision is to stop the algorithm. By applying in reverse the rule 6.36 from $m = m^* - 1$ to the first step one can determine the optimal strategy for all the pairs (m'', w), $m'' < m^*$. The results can be stored in a table and can be used later as optimal stopping rules which are *independent of the objective function.*

A stopping rule is *one-step-look-ahead subobtimal* if it stops immediately after the inequality 6.37 is true. A suboptimal rule can be used when the optimal sequential stopping rule does not exist.

Example for L_1:

$$\begin{aligned} &\mathrm{E}\left(\mathrm{E}(L_1|(m+1,\underline{w}))|(m,w)\right) - \mathrm{E}(L_1|(m,w)) = \\ &\qquad c_1^e - c_1^t\,\frac{w(w+1)}{m(m-1)}. \end{aligned} \tag{6.38}$$

This result can be obtained by substituting Formula 6.34 in Formula 6.36 (Boender and Rinnooy Kan showed it for $c_1^e = 1$, see [32]). In real problems often $c_1^t > c_1^e$, so in many cases there is no m^* for which for all the pairs (m', w) where $m' \geq m^*$ the difference 6.38 is positive (we assumed earlier that $m' \geq w+2$). Hence, the optimal sequential stopping rules for L_1 do not exist. The suboptimal rule can be applied. The algorithm should stop if

$$c_1^e - c_1^t\frac{w(w+1)}{m(m-1)} > 0. \tag{6.39}$$

Example for L_2:

Boender and Rinnooy Kan showed (see [32]) for L_2 that $m^* = \frac{c_2^t}{3}$ when $c_2^e = 1$. After an easy modification for the current case we obtain:

$$m^* = \frac{c_2^t}{3c_2^e}. \tag{6.40}$$

By applying in reverse the rule 6.36 one can determine the optimal strategy for all the pairs (m'', w), $m'' < m^*$.

6.2.3 Stopping rules that use values of the objective function

Piccioni and Ramponi (see [127] and also [83]) extended the above approach by a mechanism which takes into consideration function values in local minimizers. The stopping rules consider undiscovered minimizers in which the objective value is less than the smallest value found earlier. The following stopping rules are analogous to rules 6.29, 6.30, 6.31.

- Stop if the expected number of undiscovered *better* local minimizers is equal to zero, i.e.:

$$\mathrm{Int}\left(\frac{w}{m-w-2}\right) = 0. \tag{6.41}$$

- Stop if the expected value of the sum of relative volumes of the basins of attraction of *better* local minimizers is less than a constant η_1

$$\frac{w}{m(m-1)} < \eta_1. \tag{6.42}$$

- Stop if the probability that a *better* local minimizer does not exist is greater than a constant η_2

$$\frac{m-w-1}{m-1} < \eta_2. \tag{6.43}$$

A suboptimal stopping rule can also be estimated for L_1 in a similar way (see [83]):

- Stop if

$$c_1^e - c_1^t \frac{w}{m(m-w-2)} > 0. \tag{6.44}$$

The value of m^* for the optimal stopping rule related to the loss function L_2 can be estimated from the formula:

$$m^* = \sqrt{\frac{c_2^t}{c_2^e}}. \tag{6.45}$$

6.3 Two-phase genetic methods

6.3.1 The idea of Clustered Genetic Search (CGS)

The idea of using genetic algorithms (GA) in clustering methods in global optimization follows from the observation that GA constitute systems that transform measures (see Section 4.2). This indicates that GA should be used in order to obtain information about certain sets of positive measure rather than to obtain information about the precise location of minimizers (maximizers).

Moreover, for some types of global optimization problems, a clustering algorithm that uses GA would give not only local minimizers (maximizers) but also approximations of the basins of attraction (or some level sets which are central parts of basins) of local minimizers (maximizers). Such approximations can be helpful in many cases, for instance in parameter inverse problems, when some technical reasons imply that multi-criteria optimization methods have to be employed. One example is the problem of how to determine parameters of materials that have to be used in a construction, when parameters that correspond to exact local minimizers are not available and the choice must be delimited to some series. The essential question is how far one can go from local extremes.

In this chapter, we assume that the basins of attraction are approximated by unions of small hypercubes, i.e. raster is defined in the domain of searches. The volume of hypercubes should be a compromise between the accuracy of approximation and the capabilities of the hardware. By a cluster now we mean not the set of points that belong to the basin of attraction of a local minimizer (like in clustering methods described in Section 6.1), but a set of hypercubes (raster cells) that contain sample points.

The Clustered Genetic Search (CGS) algorithm proposed in this chapter makes use of the fact that a population of GA concentrates in the basins of attraction of local minimizers (maximizers). A rough sketch of CGS is as follows. Clusters (which are now unions of hypercubes) are recognized in a stepwise manner. Each step (a stage) results in the determination of some parts of clusters. These parts can be called *subclusters*. In each stage a genetic algorithm is started from the uniform initial population. In each stage, after the stopping rule is satisfied, the final population concentrates on parts of the basins of attraction of some minimizers. The final population determines subclusters, which are recognized by means of density analysis. Clusters are built as unions of subclusters. In subsequent stages clusters can be enlarged by attaching new subclusters. The seed of a subcluster can be the raster cell that contains "the best" individual (point). Another possibility is the cell that contains the result of a local method started from the best point. During the subcluster's recognition phase, neighboring cells, in which the density of individuals is greater than a certain threshold, are attached to the subcluster. A rough local method is started from each subcluster. It enables subclusters to join and to create clusters. The fitness function is modified on those hypercubes that are assigned to subclusters. The modification consists of setting the fitness value to the maximum that is found so far (or to an arbitrary value). The goal of this operation is to push individuals in new generations away from subclusters that are already recognized.

The proposed strategy utilizes the Simple Genetic Algorithm with the standard binary encoding, one-point cross-over and a non-zero mutation. This allows us to evaluate some properties of the whole strategy.

A simple parallel version of CGS has been tested. In this version the domain of searches is divided into subdomains. Each subdomain is divided into hypercubes of the same size. All computations are coordinated by the central coordinator called *the Master* process. *Slave* processes are distributed in a computer network. The slaves recognize subclusters in parallel. Each slave is responsible for its own subdomain.

After the global stopping rule is satisfied in each subdomain, slaves send to the coordinator the following data: minimum of the objective function and the minimizer for each cluster recognized, the size of each cluster (the number of raster cells). Exact information about each subcluster is distributed throughout the nodes. The coordinator finally joins subclusters into clusters if the distance between the subclusters' minimizers is less than the diagonal of the raster cell. The exact assignment of raster cells to clusters can be stored in the distributed manner or can be stored centrally in the coordinator (if it is possible and required).

6.3.2 Description of the algorithm

The parallel distributed version of the Clustered Genetic Search algorithm will be presented in this section.

Master

1: Divide the domain of searches $\mathcal{D}$ into p subdomains.
2: Start p slave processes.
3: Wait for the results (local minima and minimizers, sizes of subclusters).
4: After all results are obtained, join these subclusters in which the distance between minimizers is less than the diagonal of the raster cell.

Algorithm 8: Parallel version of CGS, Master.

The number of subdomains p can be greater than the number of workstations. In such a case new slave processes can be started after one of the working slaves finishes its work.

Slaves

In the version of CGS that was tested on 2-6 dimensional problems the raster was implemented in a table. Each raster cell corresponded to a cell in the table. Each table cell contained either the subcluster's number or zero (if the corresponding raster cell was not assigned). Additionally, a list of subclusters contained minimizers and minima that were found.

Figure 6.1 presents the idea of fitness modification.

A single cluster is an approximation of a central part of the basin of attraction of a local minimizer. A cluster can be recognized in one or more iterations of the outer loop (line 3 in Algorithm 9). In the inner loop (line 8) a Simple Genetic Algorithm is performed. This loop stops when the criterion

1: Do in parallel (in subdomains):
2: Determine the raster in the subdomain.
3: **repeat**
4: Generate the initial population according to the uniform distribution on the domain.
5: Evaluate fitness function $f = \tilde{\Phi}$ outside recognized subclusters.
6: Determine MAX (the largest value of $\tilde{\Phi}$).
7: Modify the fitness function ($f \leftarrow MAX$) in raster cells that are assigned to subclusters.
8: **repeat**
9: Steps of Genetic Algorithm, evaluation of subsequent generations.
10: Every certain number of generations check if subclusters can be recognized.
11: **if** Subclusters can be recognized **then**
12: Recognize subclusters using density analysis.
13: Start a local search in each subcluster.
14: Join subclusters using information from local searches (line 13).
15: **else**
16: Check if the distribution of individuals outside recognized subclusters is uniform
17: **end if**
18: **until** Complex stopping rule has been satisfied:
19: Subclusters can be recognized OR
20: The distribution of individuals outside of recognized subclusters is uniform.
21: **until**
22: Criterion from line 20 has been satisfied OR
23: All raster cells are assigned to clusters OR
24: Satisfactory set of clusters has been found.

Algorithm 9: Parallel version of CGS, slaves.

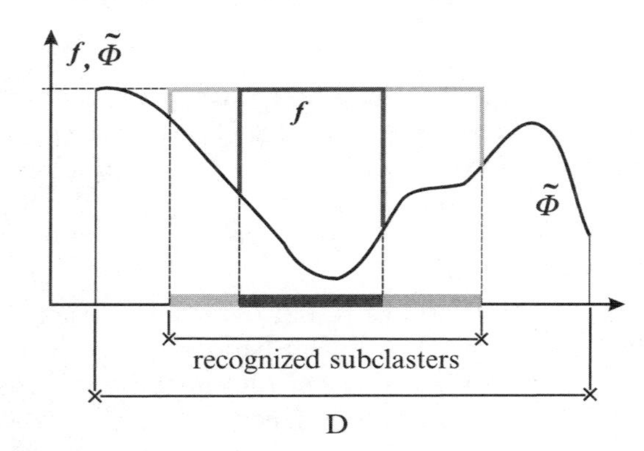

Fig. 6.1. Modification of fitness function in recognized clusters.

from line 19 is satisfied. Then the subcluster recognition process is started. The analysis of the density of individuals in raster cells is used in this process. The cell that contains "the best" individual which is not assigned to a cluster, becomes the seed of a subcluster. In the alternative tested version the seed is the raster cell that contains the minimizer found by a local optimization method started from the best individual. Neighboring cells that contain more individuals than a certain threshold are added to the subcluster. Neighboring cells are those cells that are in contact with each other. A rough local method is started in each new subcluster. If the resulting minimizer is contained in an already recognized subcluster, subclusters are joined. This can be seen in Figure 6.2.

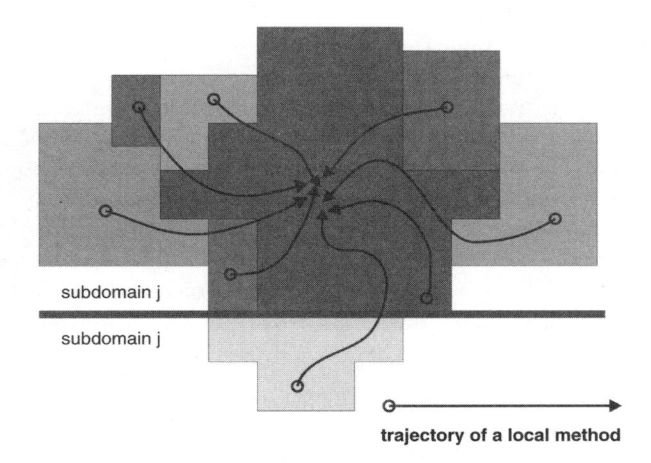

Fig. 6.2. Joining of subclusters.

6.3.3 Stopping rules and asymptotic properties

Bayesian stopping rules for clustering methods in global optimization (see Section 6.2) cannot be simply transferred to Clustered Genetic Search. The reason is the assumption made for Bayesian rules that the sample should be uniformly distributed in the domain of searches, so that the sample can easily approximate the Lebesgue measure. In CGS the distribution of individuals is usually not uniform. Moreover, it is difficult to estimate this distribution in subsequent generations. The Clustered Genetic Search algorithm differs significantly from clustering methods derived from the Multistart. Hence, different methods of analysis are required. The algorithm is closer to methods proposed by Törn. However, unlike those methods, in which after the concentration phase one can say hardly anything about the sample, for CGS we can characterize populations and some properties of the algorithm.

The stopping rule of the inner loop is complex and takes into consideration three kinds of algorithm behavior:

1. First, when the SGA finds subclusters and distinct concentrations of individuals can be determined.
2. Second, when individuals are evenly distributed outside recognized clusters. This means that an area which resembles a plateau is recognized.
3. Third, when one cannot detect any stabilization of populations after an appropriately large number of generations. This indicates that the algorithm should be stopped and the parameters of the SGA should be modified. For the sake of simplicity this is not described in Algorithm 9.

We assume that the algorithm is well tuned to the problem (see Definition 4.63), which means the limit sampling measure given by the stationary point of the genetic operator G (see Definition 4.16) concentrates on the central parts of the basins of attractions. Before the process of cluster recognition starts it is important to verify the convergence of a sequence of counting measures (given by populations) to the limit sampling measure. Any sequence of counting measures converges when $\mu \to \infty$, $k \to \infty$ (it results from Theorem 4.65). Theorem 4.66 implies that the sequence of estimators of measure density converges. Hence, a heuristic criterion which detects a stagnation of the sequence of density estimators is proposed. The criterion in line 10 in Algorithm 9 can be as follows:

- Check the convergence of the sequence of estimators of measure density. This can be treated as the "convergence" of generations. If the convergence is detected, then check if an appropriate number (given as a parameter) of raster cells contain more individuals than a certain threshold. If yes, clusters can be recognized.

Theorem 4.54 enables us to justify the stopping of the algorithm when populations of the SGA stabilize with the distribution close to uniform. This is related to the recognition of a plateau-like area (an area with a small variability of the fitness function) outside recognized clusters. If, in the subsequent generations, the SGA does not go away from a certain neighborhood of the central point of Λ^{r-1}, this point is recognized as a stationary point of the genetic operator G.

This stopping rule makes the algorithm especially useful for a certain class of objective functions, those with large plateau-like areas.

Theorem 4.54 and the assumptions imply a certain kind of asymptotic probabilistic guarantee of success in the sense that all "essential" extremes will be found. This can be justified in the following way:

- The SGA is well tuned to the problem for the set of local minimizers that belong to the set $\mathcal{W}$ (see Definition 4.63, Remark 2.2, and Remark 2.3). Theorem 4.66 guarantees that sampling measures corresponding to

populations concentrate on all the basins of attraction of local minimizers. So we expect that all essential minimizers can be discovered by clustering.
- The construction of the global stopping rule and Theorem 4.54 imply that the process of clusters recognition is not stopped before all essential minimizers are found.

What does it mean "essential" minimizer? The algorithm cannot recognize extremes with basins of attraction smaller than the raster cell. Moreover, some basins which are too shallow can be omitted because of mutation. Larger mutation rates mean that individuals fill in the basin more and more and can "overflow". Thus, some basins can "vanish", they become indistinguishable from neighboring basins. The dependence between the mutation rate and the detectability of basins is not well recognized yet and requires further studies either theoretically or practically. In tests of the proposed algorithm the values have been set after experiments.

In practice, either for optimization problems or for parameter inverse problems such a detectability condition can be justified: often only "essential" extremes are to be found, which means extremes with large and sufficiently deep basins.

The algorithm has the property of asymptotic probabilistic correctness if the global extreme is essential in the above sense.

One can see an analogy between the way in which mutation and crossover influences the proposed algorithm and the way in which the reduction phase influences such clustering methods as DC or SL. Both mechanisms mean that some extremes may stay undetected. However, unlike DC and SL with the reduction phase, the SGA is a filter which eliminates extremes that are not "essential". Moreover, the genetic algorithm can detect extremes regardless of their fitness value, but the reduction phase in classical clustering methods eliminates the possibility of detection of minima above a certain level.

The Clustered Genetic Search has certain good properties that distinguish it from other clustering methods in global optimization. The considerations presented here should be treated as the first step to further research. However, the step has its importance, because in most applications of SGA (or in general evolutionary algorithms) estimation of even asymptotic properties does not exist and heuristic stopping rules are very common.

6.3.4 Illustration of performance of Clustered Genetic Search

In this section the results of some tests of CGS for chosen functions are briefly described. The functions are: Rastrigin function, Rosenbrock function, a sine of products and a function with a large plateau-like area with two "essential" isolated local minimizers. The results for a parameter inverse problem has been presented in Chapter 2.

Plan and the goal of tests

The tests that are described in this chapter have been carried out on some known functions that are, for some reasons, difficult and are often used as test functions for methods in global optimization. In spite of the fact that they are not representative of most problems in engineering, because the cost of local methods is relatively low for them, they can be used in order to exhibit some properties of the proposed algorithm. Such purpose of the tests justifies the use of two dimensional domains. Tests for more than two dimensions have been carried out by Telega (see [184]).

The chosen test functions are sources of different difficulties for global optimizations algorithms: many local minimizers, large plateau-like areas and curved valleys. In distributed versions the domain of searches has been divided in such a way that the joining of subclusters recognized in different subdomains was needed. It was a source of an additional difficulty for the algorithm.

Each function has been tested with different parameters of the SGA engine and CGS: mutation rate, size of the population, threshold density of individuals (for the cluster recognition) and the number of generations in one iteration of the algorithm.

Rastrigin function (cosine added to a square function) has been used in order to check a filter property. Tests of this function have been carried out in two domains (a smaller and a larger one), emphasizing the different influence of two components of the sum: trigonometrical and square.

Sine of products test function has been tested with a variant of CGS without local methods started from each subcluster. This modification was necessary because the local minimizers were not isolated which was a reason of an additional difficulty.

The function with a large plateau-like area has been used in order to check usefulness of the stopping rule. In other tests the algorithm was stopped after a certain number of iterations of the outer loop. The proposed stopping rule applied to the Rastrigin or Rosenbrock function would mean that every raster cell should be assigned to some cluster and this would be ineffective.

All tests have been carried out with the use of the scalar version and the simple distributed version. The reference algorithm that was used for comparisons was a version of Multistart in which a local method is started from each raster cell. Cells for which local methods find minimizers located not further than the length of the diagonal of the cell are assigned to the same cluster. This algorithm has been called Raster Multistart (RMultistart) .

Tests made for known standard functions with different values of parameters such as the population size, stopping rule before clusters can be recognized etc., can bring some hints about how to use the algorithm for certain classes of problems. However, quantitative conclusions drawn from tests cannot be a base for definitive statements.

Remark 6.16. In figures that graphically present the results of CGS, the function map (located in the left upper part of the figure) is oriented differently than the part which shows the results of clustering. This is caused by the graphical library that was used. The same orientation can be obtained by the exchange of the X and Y axis. This remark concerns all examples. □

Rastrigin function

The function that has been tested has the following form:

$$f(x,y) = x^2 + y^2 - \cos(18x) - \cos(18y) + 2 \tag{6.46}$$

Figure 6.3 presents the two dimensional map and the three dimensional graph of the function. Different function values are presented as different shades of gray. The domain is a square $-0.5 \leq x \leq 0.5$, $-0.5 \leq y \leq 0.5$.

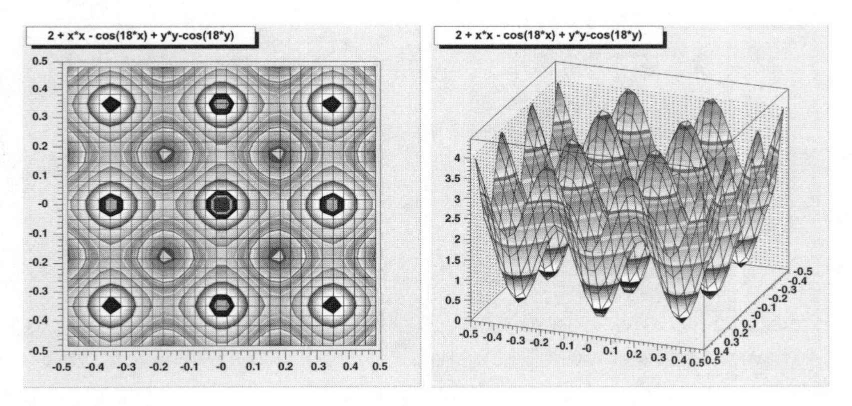

Fig. 6.3. Rastrigin function for $-0.5 \leq x \leq 0.5$, $-0.5 \leq y \leq 0.5$.

In this domain there are nine local minimizers. The global minimum equal to 0 is located at (0,0).

Figure 6.4 presents the graph of and the map of the same function but in the square domain $-10 \leq x \leq 10$, $-10 \leq y \leq 10$. minima from Figure 6.3 cannot be seen because the scale of the map.

Rastrigin Function for $-0.5 \leq x \leq 0.5$, $-0.5 \leq y \leq 0.5$

The domain is divided along x and y axis ($y = 0$, $x = 0$) into four subdomains. Each subdomain is divided into 225 raster cells (900 cells in the whole domain).

Figure 6.5 graphically presents the results of CGS for the square domain $-0.5 \leq x \leq 0.5$, $-0.5 \leq y \leq 0.5$ and the following parameters: population size $= 80$, mutation coefficient $= 0.001$, number of generations in one iteration $= 10$, number of iterations $= 20$, threshold density of individuals in one raster cell $= 12$. Parameters were not optimized. Different subclusters found in different

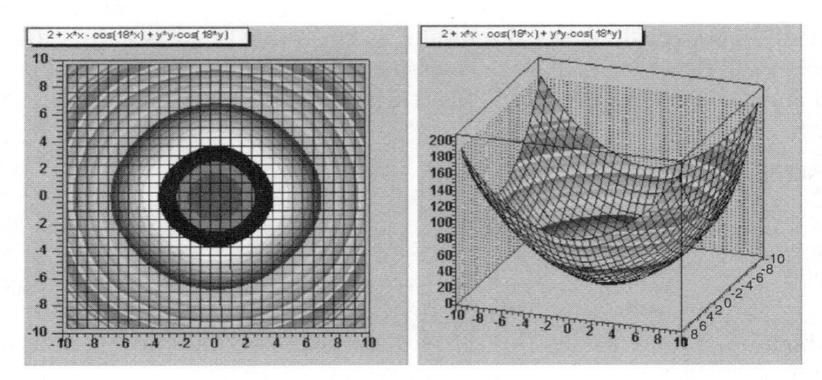

Fig. 6.4. Rastrigin function for $-10 \leq x \leq 10$, $-10 \leq y \leq 10$.

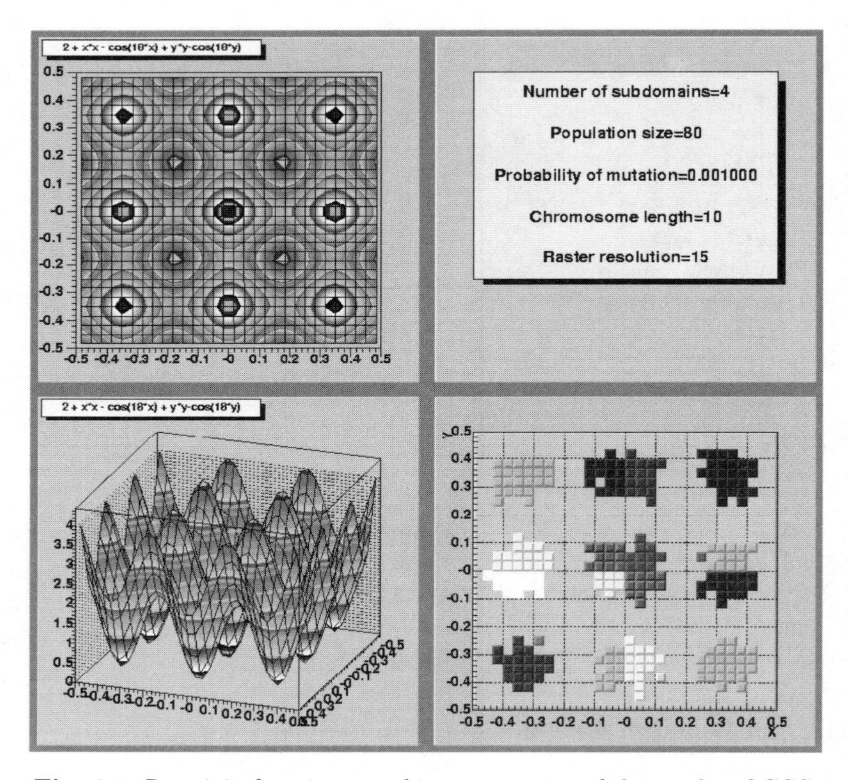

Fig. 6.5. Rastrigin function, graphic presentation of the results of CGS.

subdomains (before the coordinator joins them) are presented by different shades of gray.

All 9 local minimizers were found. The total number of objective evaluations, the number of local searches and the number of objective evaluations in each local search were counted. The number of function evaluations can be a comparative criterion with other methods. CGS was proposed for time expensive functions, so such comparison reflects the differences in execution time well. For RMultistart the number of function evaluations can be estimated as the products of the number of raster cells and the average number of function evaluations in the local search.

The number of local searches in CGS (with non-optimized parameters) was considerably less than in RMultistart (63 in comparison to 900), but the number of function evaluations was greater. The fitness function is very easy for local methods. The number of function evaluations in one local search was between 30 and 50, on average a little more than 40. The number of function evaluations in CGS was almost equal to 68000. In RMultistart it can be estimated at about 36000.

Figure 6.6 presents the results of CGS after fine tuning the parameters. The goal of fine tuning is to obtain good results of minimization with the smallest time cost of the process. The values of parameters are as follows: population size = 50, mutation rate = 0.004, number of generations in one iteration = 20, threshold density of individuals in one raster cell = 6. The number of fitness evaluations was equal to 29044, the number of local searches = 83, the number of function evaluations in local searches = 3223.

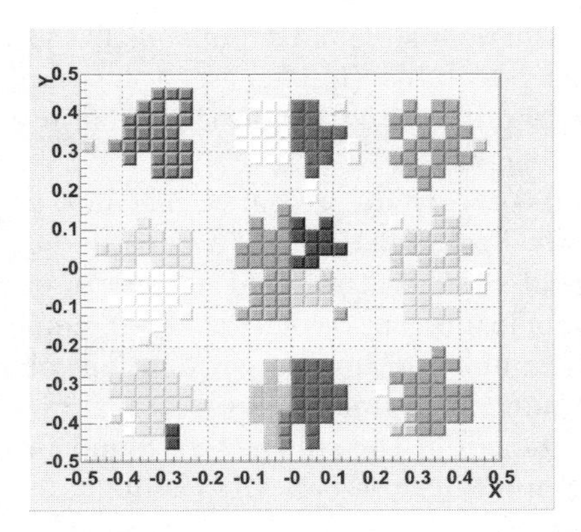

Fig. 6.6. Results of CGS after fine tuning.

In each test CGS found all 9 local minimizers in the domain. The master process joined subclusters. Nine clusters were always found.

Because the complexity of CGS can be now comparable to the complexity of RMultistart for simple ("easy") objective functions, one can expect it to be better for more "difficult" functions.

Rastrigin Function for $-10 \leq x \leq 10$, $-10 \leq y \leq 10$

Rastrigin function has many local minimizers in the square domain given by $-10 \leq x \leq 10$, $-10 \leq y \leq 10$. However, the dominant component of the sum (Formula 6.46) is the sum of the squares of x and y. When only the global minimizer is to be found, methods of smoothing of the objective function can be applied (see Coleman, Zhijun Wu [50]). The SL and MLSL algorithms are aimed at finding all local minimizers, so they are not effective for such problems.

The goal of tests of CGS for Rastrigin function in the considered domain is such a selection of parameters that cause the algorithm to "see" the objective function as the square function. Here subclusters should be related rather to level sets than to real basins of attraction of local minimizers. We want to check the following filter property: most local minimizers should remain undiscovered, the algorithm should only find the shape of the valley which contains local minimizers. Such a property can be valuable when the objective function has "noisy" valleys with many irrelevant local minimizers with small basins of attractions and similar values. In many cases, only a rough recognition of the whole valley and an approximation of the global minimizer is important. In such cases CGS can be an interesting proposition, but one modification should be made. Local searches should not be used in order to join subclusters. Clusters should be built with the use of only the density analysis. Another possibility (implemented in tests) is to join subclusters when their minimizers are located not farther than a certain threshold (for instance several cell diagonals).

Below, some results of tests are presented. The initial parameters (not optimized) were as follows: 4 subdomains, raster consisted of 900 cells, population size = 50, mutation rate = 0.02, threshold density in one cell = 4, number of generations before subclusters are recognized = 10, number of iterations = 20. The obtained results: number of function evaluations about 44000, number of local searches = 98, number of function evaluations in local searches = 4491. The master process joined all the subclusters found in subdomains into one cluster. Figure 6.7 graphically presents the results.

Figure 6.8 presents results of CGS with optimized parameters (parameters were optimized in order to obtain a lower time cost): population size = 60, mutation rate = 0.02, threshold density of individuals in one cell = 3, number of SGA generations before subclusters are recognized = 3, number of iterations = 20. The obtained results: number of function evaluations = 23797, number of local searches = 125, number of function evaluations in local searches = 5958. The master process joined all the subclusters found in subdomains into

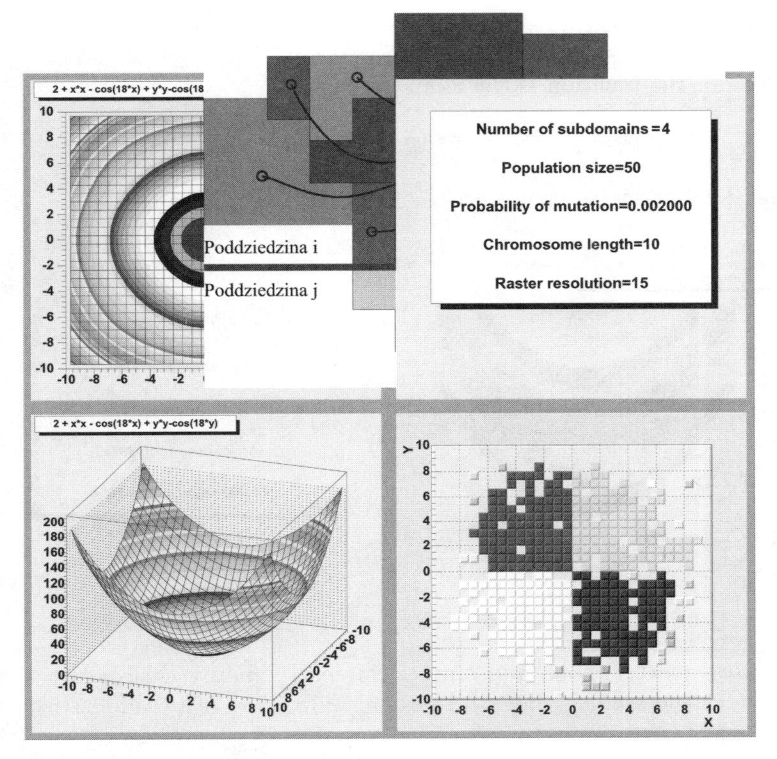

Fig. 6.7. Rastrigin function for $-10 \leq x \leq 10$, $-10 \leq y \leq 10$.

one cluster. The version without local searches was quicker: number of function evaluations = 17839.

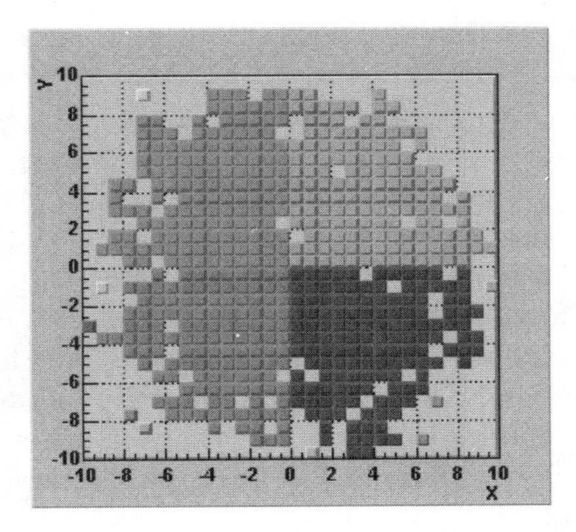

Fig. 6.8. Results of CGS after fine tuning.

Rosenbrock function

The form of the function is the following:

$$f(x,y) = 100(y - x^2)^2 + (1 - x)^2 \tag{6.47}$$

The domain was a square $-1 \leq x \leq 1$, $-1 \leq y \leq 1$.

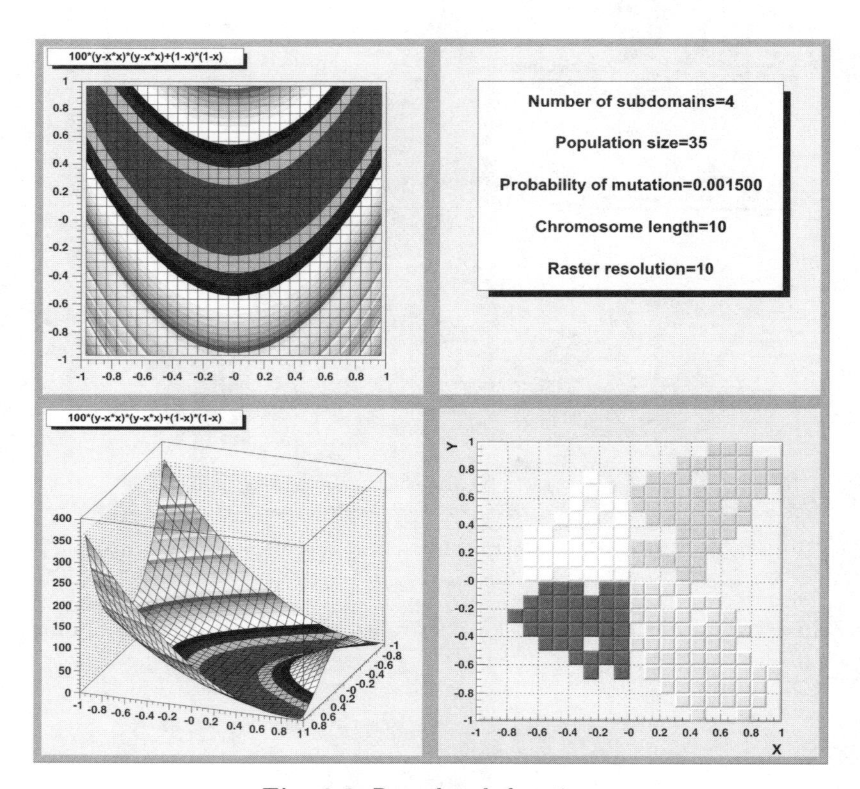

Fig. 6.9. Rosenbrock function.

This function is also known as a Rosenbrock curved valley. It has one isolated global minimizer (1,1), the minimum value is equal to 0. In the significant part of the domain the function graph is almost flat and the valley is curved. This causes difficulties for local methods. This feature is the reason why Rosenbrock function is used as a standard test function for local methods.

Figure 6.9 shows the screen with the graph of Rosenbrock Function and graphically presents the results of CGS after parameters have been refined.

The parameters of CGS were as follows: 4 subdomains, number of raster cells = 400 in the whole domain, population size = 35, mutation rate = 0.0015, threshold number of individuals in one cell = 6, number of SGA steps before subclusters are recognized = 5, number of iterations = 20. The obtained results: number of function evaluations = 19500, number of local searches = 48, number of function evaluations in local searches = 6275. The master process joined all the subclusters found in subdomains into one cluster. One global minimizer was found. In this case the average number of function evaluations in one local search was equal to 131 (in some tests even about 180). We can estimate that RMultistart would require 52400 (or 72000) function evaluations. The time cost of CGS is about 35% of the cost of RMultistart. Additionally, CGS gives more information about the basin of attraction.

Sine of a product

The form of the function is the following:

$$f(x, y) = \sin(xy) + 1 \tag{6.48}$$

The domain is a square $-3 \leq x \leq 3$, $-3 \leq y \leq 3$. This function has been tested in order to check the behavior and usefulness of CGS in cases when local minimizers are not isolated. Local methods were not used here. Subclusters were recognized only by means of analysis of the density of individuals in raster cells.

Figure 6.11 presents the function and the results of modified CGS.

The parameters were as follows: number of subdomains = 4, number of raster cells in the domain = 400, population size = 35, mutation rate = 0.01, threshold density of individuals in one raster cell = 7, number of SGA steps before subclusters are recognized = 4, number of iterations = 18. The number of function evaluations was equal to 11508. This function has been tested only with the scalar version of CGS. The conclusion of tests is that when local methods can be dispensed, better results can be obtained with the use of other methods than CGS.

A test function with large plateau-like area

Let us introduce the following notation:
(logical_condition1 AND logical_condition2)
means 1 if both logical conditions are satisfied, 0 otherwise. The form of the function is the following:

$$\begin{aligned} f(x, y) &= 12 + 0.01\sin(0.05x) + 0.01\sin(0.05y) \\ &- 0.0009(x^2 + y^2 - 10)(x > 60 \text{ AND } x < 70 \text{ AND } y > 60\text{AND } y < 70) \\ &- 0.0009(x^2 + y^2 - 10)(x > -50 \text{ AND } x < -20 \text{ AND } y > -70 \text{ AND } y < -40) \end{aligned} \tag{6.49}$$

It is the sum of the function $f_I(x, y) = 12 + 0.01\sin(0.05x) + 0.01\sin(0.05y)$ and the square function $f_{II} = -0.0009(x^2 + y^2 - 10)$ in some squares. The

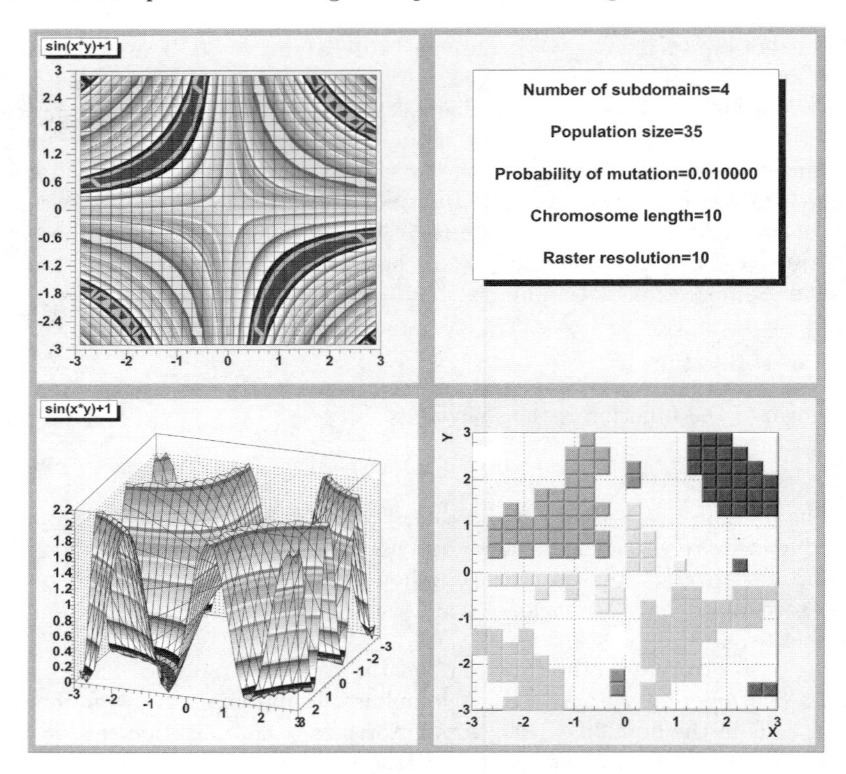

Fig. 6.10. Sine of a product.

whole domain is given by $-100 \leq x \leq 100$, $-100 \leq y \leq 100$. Figure 6.11 presents the graph of this function.

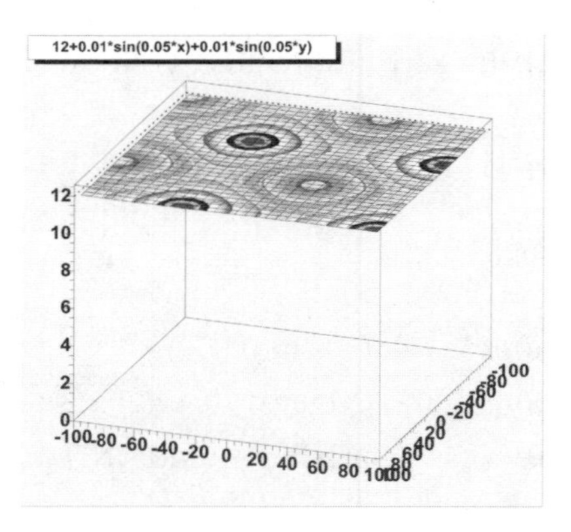

Fig. 6.11. A test function with large plateau.

"Folds" caused by the component f_I cannot be seen, because of the scale. Some tests have been carried out on the version with the stopping rule presented in Algorithm 9, some with an assumed number of iterations. Figure 6.12 presents the results of both versions (they were the same) for the following parameters (not optimized): number of subdomains = 4, number of raster cells in the whole domain = 400, population size = 40, mutation rate = 0.001, threshold number of individuals in one cell = 24, number of SGA steps before clusters can be recognized = 5.

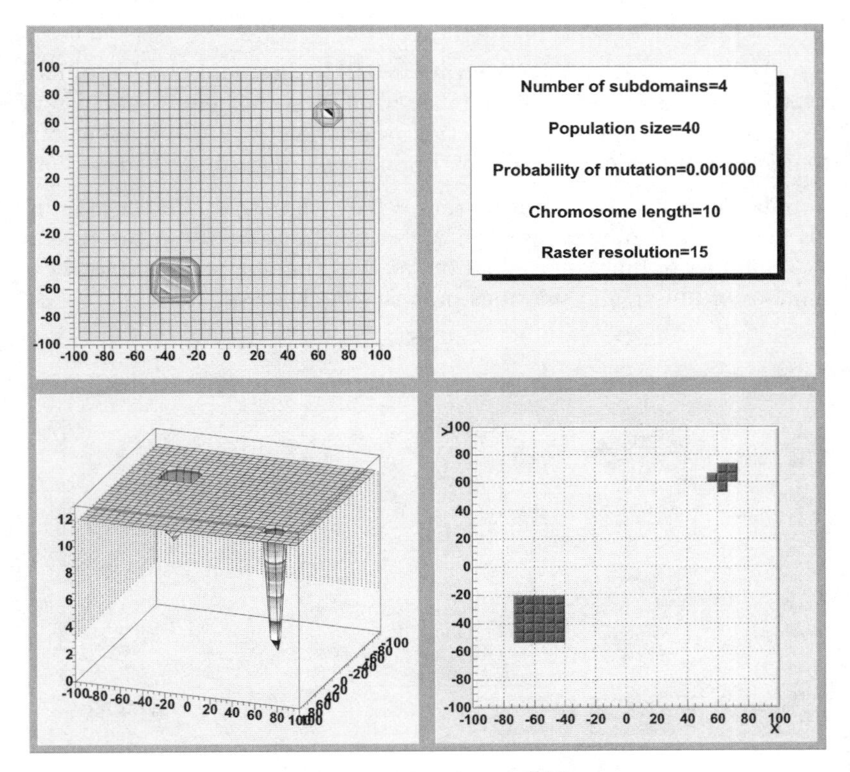

Fig. 6.12. Results of CGS.

Two clusters and two local minimizers have been found. The number of function evaluations for the version with the stopping rule given in Algorithm 9 was equal to 10306, number of local searches = 4, number of function evaluations in local searches 629. The results for the version with the simplified stopping rule were as follows: number of function calls = 19573, number of local searches = 4, number of function evaluations = 776.

The number of local searches is small. This is a good prognosis for cases when the local search is much more expensive.

The stopping rule proposed in Algorithm 9 means that the number of function evaluations is almost twice as small as for the version with the simplified stopping strategy. In two subdomains only one iteration has been performed and no local minimizer has been found. A filter property can be seen here. In fact both subdomains contained a local minimizer. Minima were omitted because their basins of attractions were too shallow to be recognized by CGS.

After the tuning of parameters the time cost was diminished, but the basins have been approximated less accurately. Figure 6.13 a) and b) presents the graphical results of CGS with the stopping rule from Algorithm 9 and the following parameters:

- population size = 20, mutation rate = 0.071, threshold number of individuals in one raster cell = 6, number of SGA steps before subclusters are recognized = 3, number of function evaluations = 2873, number of local searches = 5, number of function evaluations in local methods = 759.
- population size = 15, mutation rate = 0.09, threshold number of individuals in one cell = 6, number of SGA steps before subclusters are recognized = 3, number of function evaluations = 1967, number of local searches = 3, number of function evaluations in local searches = 652.

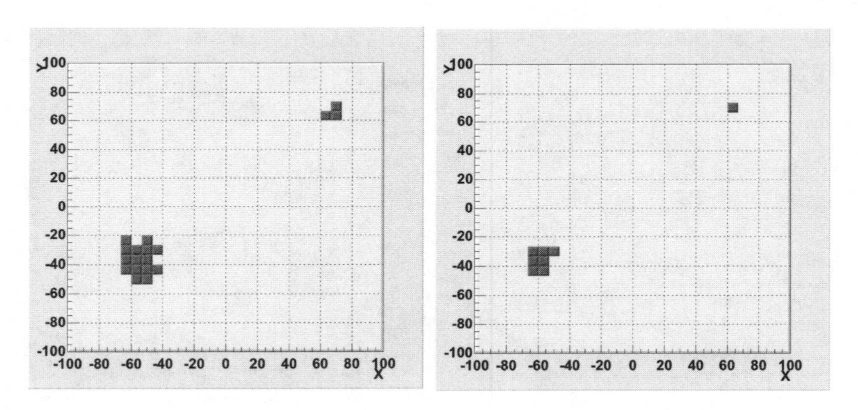

Fig. 6.13. Results of CGS for different parameters.

Summary of tests

Clustered Genetic Search can be used effectively in order to solve clustering problems in global optimization for chosen test functions. Populations concentrate in the basins of attraction of local minimizers. When parameters of the algorithm are refined, CGS can be much less expensive than the reference RMultistart. CGS can be especially effective for objective functions with large plateau-like areas. However, so far there is no other answer to the question of how parameters should be chosen than the answer they should be set via tests. This is the common way in which parameters of genetic algorithms are set in most applications. On the other hand, there are some estimations about the asymptotic properties of CGS and this singles out this method. Test showed also that CGS has some good properties that distinguish it from standard clustering methods in global optimization. These properties can be advantageous in many practical problems.

The CGS algorithm presented in this book should be treated as the first step towards using genetic algorithms in clustering global optimization methods. Further research and tests are required. Some improvements to CGS have been proposed by Telega, Podolak in [129]. The goal of modifications is to store information about basins more effectively. Ellipsoids can be used instead of hypercubic raster or, in a more robust version, clusters can be remembered by a neural network.

7

Summary and perspectives of genetic algorithms in continuous global optimization

We would like to close this book with several short remarks that summarize the research into the design and analysis of genetic algorithms, as well as their application in solving continuous global optimization problems. Some conjectures concerning the direction of further research into these areas will also be made.

Global optimization algorithms, based on genetic mechanisms inherited from biology, can be applied effectively when solving continuous problems only in difficult cases in which other, much faster computing strategies have failed. This is mainly due to the large computational cost, even if simple genetic techniques are run. Continuous global optimization problems with multimodal objective functions, in which the basins of attraction of local extremes are separated by large areas on which moderate or low objective variability is observed (plateaus), belong to the group of important difficult ones. Another type of difficulty is caused by the low regularity of the objective function when the gradient and the Hessian computation require costly approximative routines or is generally meaningless.

Genetic algorithms adapted to solve continuous problems are usually ineffective or even ill defined if we need to find a very accurate approximation of extremes. However, they may satisfactorily compete with the Monte Carlo methods as global phase algorithms in two-phase stochastic global optimization strategies.

The review of research into genetic algorithms presented in this book exhibits two of their important directions:

- Studying the population dynamics modeled as a single point in the specially defined space of states $\mathcal{E}$. The leading model in this group is the stochastic process as a dynamic system in the space of probabilistic measures $\mathcal{M}(\mathcal{E})$. For several cases (e.g. the Simple Genetic Algorithm SGA) this model may be reduced to the uniform Marcov chain. In this book, the results of this approach were mainly described in Sections 4.1.2 and 4.2.2. Moreover, the dynamics rule of the sampling measures on the admissible

R. Schaefer: *Foundation of Global Genetic Optimization*, Studies in Computational Intelligence (SCI) **74**, 199–201 (2007)
www.springerlink.com

domain $\mathcal{D}$ may be lead from this model. The results of this group may be utilized for convergence verification and the construction of stopping rules. They are especially helpful in verifying the stopping rules of genetic algorithms used at the global phase of two-phase stochastic global optimization strategies (see Section 6.3.3).

- Studying of the local behavior of the individual in an evolving population. The model of global convergence mentioned in Section 4.1.3 and the results presented in Section 4.2.1 fall into this group. In addition, important research directions, such as Building Block Theory for discrete encoding genetic algorithms as well as evaluation of the first hitting time for the neighborhood of the global extreme, may be added to this group (see e.g. Garnier, Kaller, Schoenauer [72], Rudolph [142]). Theoretical results obtained in this way require strong assumptions with respect to both the genetic algorithm and the optimization problem.

An interesting new direction of research, which is not described in this book, is the study of spacial statistic dynamics of medium size populations (e.g. mean phenotype dynamic or the dynamic of the dynamics of the standard deviation of population phenotypes). Such results may be found in the papers of Chorążyczewski and Galar [48], [46], [47]. Their results show, that medium-sized populations remain a compact group of points in the phenotype space during the whole evolution time. The global search is performed by the whole population displacement toward the consecutive basins of attraction of local extremes. This movement proceeds periodically by repeating three phases:

- the climbing phase in which the population bounds to the central part of the basin of attraction of the particular local extreme of the objective function,
- the phase of filling the central part of this basin of attraction,
- the phase of crossing the saddle in the evolutionary landscape toward the next basin.

Moreover authors motivate that there is almost impossible to design the single genetic algorithm which may behave effectively in each of three phases. Some adaptation techniques that can relax this obstacle were mentioned in the Chapter 5.

The local character of the single population genetic search, as well as the difficulty of designing a universal algorithm that can be effectively utilized in each of the three phases described above, motivate the theoretical considerations and inventions toward multi-deme parallel genetic searches. In such strategies, deme components share tasks of various accuracy degree searches, search regions and can also be specialized in climbing, filling or saddle crossing separately.

The basic multi-deme model that incorporates the above requirements is the island model (see Section 5.4.2). It seems, that hierarchical strategies described in the Section 5.4.3 satisfied these needs more comprehensively. The result of multi-deme searches may be strengthen by the proper post processing (e.g. Clustered Genetic Search described in Section 6.3) and the next phase of solving the global optimization problem by using accurate, low cost local optimization methods. The high efficiency of multi-deme global searches and their post-processing is confirmed by mathematical derivations and numerical experiments (see e.g. Whitley, Soraya, Hackerdorn [200], Kołodziej [95] Cantú-Paz [44], [45] and Schaefer, Adamska, Telega [152]).

Summing up, multi-deme genetic strategies supported by refined post-processing techniques constitute the most promising direction in the global optimization search in large, multidimensional continuous domains.

List of Symbols

Symbol	Meaning
V	solution space for the global optimization problem
$d(\cdot,\cdot)$	distance function in the space V
$d_H(\cdot,\cdot)$	Hamming distance function
$\|\cdot\|_2$	Euclidean norm in the space V
$\|\cdot\|$	arbitrary norm in the space V
N	dimension of the global optimization problem ($N = \dim(V)$)
$\mathcal{D}$	admissible set of solutions to the global optimization problem
$\Phi : \mathcal{D} \rightarrow \mathbb{R}_+$	objective function of a maximization problem
$\tilde{\Phi} : \mathcal{D} \rightarrow \mathbb{R}_+$	objective function of a minimization problem
x^*	global maximizer (or minimizer) to the global optimization problem
x^+	local maximizer (or minimizer) to the global optimization problem
$\mathrm{meas}(\cdot)$	Lesbegue measure function
Loc	set of local optimization methods
$\mathrm{loc}(\cdot)$	local optimization method
$R_{x^+}^{\mathrm{loc}}$	attractor of the local minimizer x^+ with respect to the strictly decreasing local optimization method $\mathrm{loc}(\cdot)$
$\mathcal{B}_{x^+}$	basin of attraction of the local minimizer x^+
P	random sample, population
$\Pr(\cdot)$	probability measure
$\mathcal{M}(A)$	space of probabilistic measures over the set A
$\mathrm{b}(\cdot)$	function that selects the best fitted individual in a population
$\mathcal{D}_r \subset \mathcal{D}$	grid of encoded points (set of phenotypes)
$r = \#\mathcal{D}_r$	the cardinality of phenotype set
U	genetic universum (set of genotypes)
$\mathrm{code} : U \rightarrow \mathcal{D}_r$	encoding bijective function
$\mathrm{dcode} : \mathcal{D} \longmapsto U$	partial function of inverse encoding (decoding partial function)
Ω	binary genetic universum in Chapters 3 - 5, space of elementary events in Section 6.1.5
l	binary code length
Z_2	group $\{0,1\}$ with the addition modulo 2

$\oplus$	the addition operator in Z_2, coordinate-by-coordinate addition of binary vectors from $Z_2 \times, \ldots, \times Z_2$
$\mathrm{code}_a(\cdot)$	affine binary encoding
$\mathrm{code}_G(\cdot)$	Gray binary encoding
$f : U \to \mathbb{R}_+$	fitness function
$\mathrm{Scale}(\cdot)$	nonlinear function that modifies fitness
μ	number of individuals in the parental population
$\mathrm{sel}_f(\cdot)$	probability distribution of selection
$Elite$	subset of individuals that pass to the next epoch with the probability 1
$\mathbf{1}$	vector of l ones
$\hat{i}$	the inverse of the binary vector i ($\hat{i} = \mathbf{1} \oplus i$)
p_m	rate of binary mutation
p_c	rate of binary crossover
$[\cdot]$	evaluation function for boolean expressions
$\otimes$	coordinate-by-coordinate multiplication of binary vectors
$type$	type of binary crossover
$\mathrm{mut}_x(\cdot)$	probability distribution for the binary mutation of an individual x
$\mathrm{cross}_{x,y}(\cdot)$	probability distribution for the binary crossover of individuals x and y
$\mathcal{N}(e, \mathbf{C})$	realization (result of sampling) of the N-dimensional Gauss random variable with the mean vector e and the covariance matrix $\mathbf{C}$
$\mathcal{N}(e, \sigma)$	realization (result of sampling) of the one dimensional Gauss random variable with the mean value e and the standard deviation σ
$\mathcal{U}[0, 1]$	realization (result of sampling) of the random variable of the uniform probability distribution over the real interval $[0, 1]$
λ	number of offspring individuals in a single genetic epoch
κ	individual life time parameter of the evolutionary algorithm in section 5.3.2, raster resolution in section 6.1.10
$\mathcal{S}$	space of schemata
H	schemata from the space $\mathcal{S}$
$\Delta(H)$	length of the schemata H
$\aleph(H)$	degree of the schemata H
$\mathrm{E}(\cdot)$	expected value operator
$\mathcal{E}$	space of states of genetic algorithms
eqp	equivalence relation among vectors of genotypes
S_μ	group of permutations of the μ-element set
Λ^{r-1}	unit $r-1$ simplex in $\mathbb{R}^r$
$X_\mu \subset \Lambda^{r-1}$	finite set of states of the genetic algorithm with finite population ($\mu < +\infty$)
$\#A$	cardinality of the set or multiset A
$n = \#\mathcal{E}$	cardinality of the space of states
π^t	genetic algorithm state probability distribution in the epoch t
τ	Markov transition function of genetic algorithm states
$\mathbf{Q}$	probability transition matrix
$F(\cdot)$	selection operator for the simple genetic algorithm

$M(\cdot)$	mixing operator for the simple genetic algorithm
$G(\cdot)$	genetic (heuristics) operator for the simple genetic algorithm
$\mathbf{M}$	mixing matrix
$\mathcal{K} \subset \Lambda^{r-1}$	set of fixed points of the genetic operator G
$\dim(V)$	dimension of the vector space V
$\mathrm{diam}(A)$	diameter of the subset A of a metric space
$\mathbb{N}$	set of natural numbers
$\mathbb{Z}$	set of integers
$\mathbb{Z}_+$	set of nonnegative integers $\mathbb{Z}_+ = \mathbb{N} \cup \{0\}$
$\mathbb{R}$	set of rational numbers
$\mathbb{R}_+$	set of nonnegative rational numbers
$\mathrm{top}(V)$	topology on V (the family of open sets in V)
$\overline{A}$	the topological closure of the set A in the proper topology
$(A)^-$	the complement of the set A, i.e. $(A)^- = V \setminus A$ if it is contained in the space V
χ_A	the characteristic function of the set A. Assuming A is contained in the space V $\chi_A : V \rightarrow \{0, 1\}, \chi_A(x) = 1$ if $x \in A, \chi_A(x) = 0$ otherwise
$\lceil \cdot \rceil$	operator of the upper round "ceiling"
$\lfloor \cdot \rfloor$	operator of the lower round "floor"
$\mathrm{Int} : \mathbb{R} \rightarrow \mathbb{N}$	function turning back the nearest integer to the argument
$\mathrm{int}(A)$	interior of the set A in the proper topology
$\mathrm{diag} : \mathbb{R}^N \rightarrow \mathbb{R}^{N^2}$	function turning back the square diagonal matrix $\mathrm{diag}(v)$ with the diagonal equal to the vector v
$\mathrm{supp}(g)$	support of the real valued function $g : A \rightarrow \mathbb{R}$, $\mathrm{supp}(g) = \{x \in A; g(x) \neq 0\}$
I	identity mapping
$\mathbf{I}$	the matrix of linear identity of finite dimensional vector space
$\mathrm{Dom}(f)$	domain of the function f
$u(t)$	vector of parameters that control genetic operations in the epoch t
$\mathcal{L}(\mathbb{R}^r \rightarrow \mathbb{R}^r)$	space of linear mappings from $\mathbb{R}^r$ into itself
Γ	Gamma Euler's function

References

1. Ali M, Storey C (1994) Topographical Multilevel Single Linkage. *Journal of Global Optimization* 5:349–358.
2. Anderson RW (1997) The Baldwin Effect. Chapter C.3.4.1 in [15].
3. Anderssen RS, Bloomfield P (1975) Properties of the random search in global optimization. *Journal of Optimization Theory and Applications* 16:383–389.
4. Arabas J (1995) *Evolutionary Algorithms with the varying population cardinality and the varying crossover range.* PhD Thesis, Warsaw University of Technology, Warsaw (in Polish).
5. Arabas J (2001) *Lectures in Evolutionary Algorithms.* WNT, Warsaw (in Polish).
6. Arabas J (2003) Sampling measure of an evolutionary algorithm. In: *Proc. of 6-th KAEiOG Conf.*, Łagów Lubuski, 15–20.
7. Arabas J, Michalewicz Z, Mulawka J (1994) GAVaPS – a Genetic Algorithm with Varying Population Size. *Proc. of ICEC'94*, Orlando, Florida, IEEE Press, 73–76.
8. Arabas J, Słomka M (2000) Pseudo-random number generators in the initial population generation. In: *Proc. of 4-th KAEiOG Conf.*, Lądek Zdrój, 7–12 (in Polish).
9. Archetti Betrò B (1978) *On the Effectiveness of Uniform Random Sampling in Global Optimization Problems.* Technical Report, University of Pisa.
10. Bäck T, Fogel DB, Michalewicz Z eds. (2000) *Evolutionary Computation 1. Basic Algorithms and Operators.* Institute of Physics Publishing, Bristol and Philadelphia.
11. Bäck T, Fogel DB, Michalewicz Z eds. (2000) *Evolutionary Computation 2. Advanced Algorithms and Operators.* Institute of Physics Publishing, Bristol and Philadelphia.
12. Bäck T (1996) *Evolutionary Algorithms in Theory and Practice.* Oxford Univ. Press.
13. Bäck T (1997) Self–adaptation. Chapter C.7.1 in [15].
14. Bäck T (1997) Mutation Parameters. Chapter E.1.2 in [15].
15. Bäck T, Fogel DB, Michalewicz Z (1997) *Handbook of Evolutionary Computations.* Oxford University Press.
16. Bäck T, Schütz M (1996) Intelligent Mutation Note Control in Canonical Genetic Algorithm. Foundation of Intelligent Systems. In: Ras ZW, Michalewicz Z eds. *Proc. of 9-th Int. Symp. ISIM'96.*, *LNCS* 1079, Springer.

17. Bagley JD (1967) *The Behavior of Adaptive Systems with Employ Genetic and Correlation Algorithms.* PhD Thesis, University of Michigan, *Dissertation Abstracts International* 28(12), 5106B. (Univ. Microfilms No. 68–7556).
18. Bahadur RR (1966) A Note on Quantiles in Large Samples. *Annals of Mathematical Statistics* 37:577–580.
19. Beasley D, Bull DR, Martin RR (1993) A Sequential Niche for Multimodal Function Optimization. *Evolutionary Computation* Vol. 1, No. 2, 101–125.
20. Becker RW, Lago GV (1970) A Global Optimization Algorithm. In: *Proc. of Allerton Conf. on Circuits and System Theory,* Monticallo, Illinois, 3–15.
21. Bethke AD (1981) *Genetic Algorithms as Function Optimizers.* PhD Thesis, University of Michigan, *Dissertation Abstracts International* 41(9), 3503B, (Univ. Microfilms No. 8106101).
22. Betrò B (1981) *Bayesian Testing of Nonparametric Hypotheses and its Application to Global Optimization.* Technical Reports CNR-IAMI, Italy.
23. Betrò, B Schoen F(1987) Sequential Stopping Rules for the Multistart Algorithm in Global Optimization. *Mathematical Programming* 38:271–286.
24. Betrò, B Schoen F(1992) Sequential Stopping Rules for the Multistart Algorithm in Global Optimization. *Mathematical Programming* 52:445–458.
25. Beyer HG (1995) Toward a Theory of Evolution Strategies: Self-adoption. *Evolutionary Computation* 3:311–348.
26. Beyer HG (2001) *The Theory of Evolution Strategies.* Springer.
27. Beyer HG, Rudolph G (1997) Local Performance Measures. Chapter B.2.4. in [15].
28. Billingsley P (1979) *Probability and Measure.* John Willey and Sons, New York, Chichester, Brisbane, Toronto.
29. Boender CGE (1984) *The Generalized Multinominal Distribution: A Bayesian Analysis and Applications.* PhD Thesis, Erasmus University, Rotterdam, Centrum voor Wiskunde en Informatica, Amsterdam.
30. Boender CGE, Rinnooy Kan AHG (1991) On when to Stop Sampling for the Maximum. *Journal of Global Optimization* 1:331–340.
31. Boender CGE, Rinnoy Kan AHG (1985) *Bayesian Stopping Rules for a Class of Stochastic Global Optimization Methods.* Technical Report, Econometric Institute, Erasmus University, Rotterdam.
32. Boender CGE, Rinnoy Kan AHG (1987) Bayesian Stopping Rules for Multistart Global Optimization Methods. *Mathematical Programming* 37:59–80.
33. Boender CGE, Rinnoy Kan AHG, Vercellis C (1987) Stochastic Optimization. In: Andreatta G, Mason F, Serafini P eds. *Advanced School on Stochastics in Combinatorial Optimization.* Word Scientific, Singapore.
34. Boender CGE, Rinnoy Kan AHG, Stougie L, Timmer GT (1982) A Stochastic Method for Global Optimization. *Mathematical Programming* 22:125–140.
35. Boender CGE, Zieliński R (1985) A Sequential Bayesian Approach to Estimating the Dimension of a Multinominal Distribution. In: *Sequential Methods in Statistics. Banach Center Publications* Vol. 16, PWN–Polish Scientific Publisher, Warsaw.
36. Borowkow AA (1972) *A course in probabilistic.* Nauka, Moscow (in Russian).
37. Brooks SH (1958) A Discussion of Random Methods for Seeking Maxima. *Operational Research* 6:244–251.
38. Burczyński T (1955) *Boundary Element Method in Mechanics.* WNT, Warsaw (in Polish).

39. Burczyński T, Długosz A, Kuś W, Orantek P (2000) Evolutionary Design in Computer Aided Engineering. In: *Proc. of 5-th Int. Conf. on Computer Aided Engineering*, Polanica Zdrój.
40. Burczyński T, Kuś W, Nowakowski M, Orantek P (2001) Evolutionary Algorithms in Nondestructive Identification of Internal Defects. In: *Proc. of 5-th KEGiOG Conf.*, Jastrzębia Góra, 48–55.
41. Burczyński T, Orantek P (1999) The hybrid genetic–gradient algorithm. In: *Proc. of 3-rd KEGiOG Conf.*, Potok Złoty, 47–54 (in Polish).
42. Cabib E, Davini C, Chong-Quing Ru (1990) A Problem in the Optimal Design of Networks Under Transverse Loading. *Quaternary of Appl. Math.* Vol. XLVIII, 252–263.
43. Cabib E, Schaefer R, Telega H (1998) A Parallel Genetic Clustering for Inverse Problems. *LNCS* 1541, Springer, 551–556.
44. Cantú Paz E (2000) Markov Chain of Parallel Genetic Algorithm. *IEEE Transactions on Evolutionary Computation* 4:216–226.
45. Cantú Paz E (2000) *Efficient and accurate parallel genetic algorithms.* Kluwer Academis Publishers.
46. Chorążyczewski A (2000) Some Restrictions of the Standard Deviation Modifications in Evolutionary Algorithms. In: *Proc. of 4-th KEGiOG Conf.*, Lądek Zdrój, 45–50 (in Polish).
47. Chorążyczewski A (2001) *The Analysis of the Adoption Skills of Evolving Population and their Applications in Global Optimization.* PhD Thesis, Wrocław University of Technology, Wrocław (in Polish).
48. Chorążyczewski A, Galar R (2001) Evolutionary Dynamics in Space of States. In: *Proc. of CEC'2001* Seoul, Korea, 1366–1373.
49. Chow YS, Teicher H (1978) *Probability Theory.* Springer.
50. Coleman TF, Zhijun Wu (1995) Parallel continuation-based global optimization for molecular conformation and protein folding. *Journal of Global Optimization* Vol. 8, No. 1, 49–65.
51. Cromen TH, Leiserson ChE, Rivest RL, Stein C (2001) *Introduction to Algorithms.* MIT Press.
52. Davis L (1989) Adapting Operator Probabilities in Genetic Search Algorithms. In: *Proc. of ICGA'89 Conf.*, San Mateo, CA, Morgan Kaufman.
53. Davis L ed. (1991) *Handbook of Genetic Algorithms.* Van Nostrand Reinhold, New York.
54. De Jong KA (1975) *An Analysis of the Behavior of a Class of Genetic Adaptive Systems.* PhD Thesis, Univ. of Michigan.
55. De Jong KA, Sarma J (1995) On Decentralizing Selection Algorithm. In: Eselman LJ ed. *Proc. of 6-th Conf. on Genetic Algorithms*, Morgan Kaufman, 17–23.
56. Deb K (2000) Encoding and Decoding Functions. Chapter 2 in [11].
57. Deb K, Goldberg D (1989) An investigation of Niche and Species Formation in Genetic Function Optimization. In: Schaffer JD ed. *Proc. of Int. Conf. on Genetic Algorithms*, Morgan Kaufman, 42–50.
58. Derski W (1975) *An Introduction to the Mechanics of Continua.* PWN, Warsaw (in Polish).
59. Devroye L (1978) Progressive Global Random Search of Continuous Function. *Mathematical Programming* 15:330–342.
60. Dixon LCW, Szegö GP eds. (1975) *Toward Global Optimization.* North Holland, Amsterdam.

61. Dixon LCW, Gomulka J, Szegö GP (1975) Toward Global Optimization. In [60].
62. Dulęba I, Karcz–Dulęba I (1996) The Analysis of a Discrete Dynamic System Generated by an Evolutionary Process. In: *Works of IX Symposium: Simulation of Dynamic Processes*, Poland, 351–356 (in Polish).
63. Eiben AE, Raué PE, Ruttkay Z (1994) Genetic Algorithms with Multi-Parent Recombination. In: Davidor Y, Schwefel HP, Mäner R *Proc. of PPSN III, Lecture Notes in Computer Science* 866, Springer:68–77.
64. Eshelman L (1991) The CHC Adaptive Search Algorithm: How to have Safe Search when Engaging in Nontraditional Genetic Recombination. In: Rawlins G ed. *Foundations of Genetic Algorithms*, Morgan Kaufman, 265–283.
65. Feller W (1968) *An Introduction to Probability. Theory and Applications.* John Wiley & Sons Publishers.
66. Findeisen W, Szymanowski J, Wierzbicki A (1980) *Theory and Methods of Optimization.* PWM, Warsaw (in Polish).
67. Fogarty TC (1989) Varying the Probability of Mutation in the Genetic Algorithm. In: Schaffer JD ed. *Proc. of 3-rd Int. Conf. on Genetic Algorithms*, San Mateo CA, Morgan Kaufman, 104–109.
68. Fogel TC (1992) *Evolving Artificial Intelligence.* PhD Thesis, Univ. of California.
69. Freisleben B (1997) Metaevolutionary Approaches. Chapter C.7.2 in [15].
70. Galar R, Karcz–Dulęba I (1994) The Evolution of Two. An Example of Space of States Approach. In: Sebald AV, Fogel LJ eds. *Proc. of the Thrid Annual Conf. on Evolutionary Programming*, San Diego CA, Word Scientific, 261–268.
71. Galar R, Kopcouch R (1999) Ipaciency and Polarization in Evolutionary Processes. In: *Proc. of 3-rd KAEiOG Conf.*, Potok Złoty, 115–122 (in Polish).
72. Garnier J, Kaller L, Schoenauer M (1999) Rigorous Hitting Times for Binary Mutation. *Evolutionary Computation* Vol. 7, No. 2, 167–203.
73. Goldberg D, Richardson J (1987) Genetic Algorithms with Sharing for Multimodal Function Optimization. Genetic Algorithms and Their Applications. In: *Proc. of 2-nd Int. Conf. on Genetic Algorithms*, Lawrence Erlbaum Associates, Inc., 41–49.
74. Goldberg DE (1989) *Genetic Algorithms and their Applications.* Addison–Wesley.
75. Goldberg D (1989) Sizing Populations for Serial and Parallel Genetic Algorithms. *Proc. Third Int. Conf. on Genetic Algorithms*, Morgan Kaufman, 70–79.
76. Goldberg DE, Deb K, Clark J (1992) Genetic Algorithms, Noise and the Sizing of Populations. *Complex Systems* 6:333–362.
77. Greffenstette JJ (1986) Optimization of Control Parameters of Genetic Algorithms. *IEEE Transactions on Systems, Man and Cybernetics* Vol. SMC-16, No. 1.
78. Greffenstette JJ (1993) Deception Considered Harmful. In: Rawlins G ed. *Foundations of Genetic Algorithms*, Morgan Kaufman.
79. Greffenstette JJ, Bayer JE (1989) How Genetic Algorithms Work. A Critical Look at Implicit Parallelism. In: Schaffer A ed. *Proc. of the 3-rd Int. Conf. on Genetic Algorithms*, Morgan Kaufman.
80. Grygiel K (1996) On Asymptotic Properties of a Selection-with-Mutation Operator. In: *Proc. of 1-th KAEiOG Conf.*, Murzasichle, Poland, 50–56.

81. Grygiel K (200) Genetic algorithms with AB-mutation. In: *Proc. of 4-th KAEiOG Conf.*, Lądek Zdrój, Poland, 91–98 (in Polish).
82. Grygiel K (2001) Mathematical models on Evolutionary Algorithms. In: Schaefer R, Sędziwy S eds. *Advances in Multi-Agent Systems.* Jagiellonian University Press, Kraków, 139–148.
83. Guus C, Boender E, Romeijn EH (1995) Stochastic Methods. In [86].
84. Harik G, Cantú-Paz E, Goldberg DE, Miller BL (1997) The Gambler's Ruin Problem, Genetic Algorithms and Sizing Populations. In: Bäck T ed. *Proc. of the IEEE Int. Conf. on Evolutionary Computation*, Piscataway, NJ, USA:IEEE, 7–12.
85. Holland JH (1975) *Adaptation in Natural and Artificial Systems.* Univ. of Michigan Press, Ann. Arbor.
86. Horst R, Pardalos PM (1995) *Handbook of Global Optimization.* Kluwer.
87. Hulin M (1997) An Optimal Stop Criterion for Genetic Algorithm, a Bayesian Approach. In: Bäck T ed. *Proc. of ICGA'97*, 135–142.
88. Iosifescu M (1988) *Finite Markov Processes and their Applications.* PWN, Warsaw (in Polish).
89. Jain AK, Murty MN, Flynn PJ (1999) Data Clustering *ACM Computing Surveys*, Vol. 31, No 3, 264–323.
90. Julstrom BA (1995) What Have You Done for Me Lately? Adapting Operator Probabilities in a Steady-State Genetic Algorithm. In: Eshelman LJ ed. *Proc. of ICGA'95*, Pittsburg, Pennsylvania, Morgan Kaufman, 81–87.
91. Karcz-Dulęba I (2000) The Dynamics of Two-Element Population in the State Space. The Case of Symmetric Objective Functions. In: *Proc. of 4-th KAEiOG Conf.*, Lądek Zdrój, 115–122 (in Polish).
92. Karcz-Dulęba I (2001) Evolution of Two-Element Population in the Space of Population States. Equilibrium States for Asymetrical Fitness Functions. In: *Proc. of 5-th KAEiOG Conf.*, Jastrzębia Góra, 106–113.
93. Kieś P, Michalewicz Z (2000) Foundations of Genetic Algorithms. *Matematyka Stosowana* 1:68–91 (in Polish).
94. Kołodziej J (1999) Asymptotic behavior of Simple Genetic Algorithm. In: *Proc. of 3-rd KAEiOG Conf.*, Potok Złoty, 167–174.
95. Kołodziej J (2001) Modeling Hierarchical Genetic Strategy as a Family of Markov Chains. In: *Proc. of 4-th PPAM Conf.*, Nałęczów, Poland, *LNCS* 2328, Springer, 595–598.
96. Kołodziej J (2003) Hierarchical Strategies of the Genetic Global Optimization. PhD Thesis, Jagiellonian University, Faculty of Mathematics and Informatics, Kraków, Poland (in Polish).
97. Kołodziej J, Schaefer R, Paszyńska A (2004) Hierarchical Genetic Computation in Optimal Design. *Journal of Theoretical and Applied Mechanics*, Vol. 42, No. 3, 78–97.
98. Kołodziej J, Jakubiec W, Starczak M, Schaefer R (2004) Identification of the CMM Parametric Errors by Hierarchical Genetic Strategy Applied. In: Burczyński T, Osyczka A eds. *Solid mechanics and its Applications*, Vol. 117, Kluwer, 187–196.
99. Koza JR (1992) *Genetic Programming. Part 1, 2.* MIT Press, (1992–Part 1, 1994–Part 2).
100. Krishnakumar K (1989) Micro-Genetic Algorithms for Stationary and Non-stationary Function Optimization. In: *SPIE's Intelligent Control and Adaptive Systems Conf.*, Vol. 1196, Philadelphia, PA.

101. Kwietniak D (2006) Chaos in the Devaney sense, its variants and topological entropy. PhD Thesis, Jagiellonian University, Faculty of Mathematics and Informatics, Kraków, Poland (in Polish).
102. Langdon WB, Poli R (2002) *Foundation of Genetic Programming.* Springer.
103. Lis J (1994) *Classification algorithms based on Artificial Neural Networks.* PhD thesis, Inst. of Biocybernetics and Biomedical Eng., Polish Academy of Science, Warsaw (in Polish).
104. Lis J (1995) Genetic Algorithms with the Dynamic Probability of Mutation in the Classification Problem. *Pattern Recognition Letters* 16:1311–1321.
105. Littman M, Ackley D (1991) Adaptation in Constant Utility Nonstationary Environment. In: Belew R, Broker L eds. *Proc. of ICGA'91 Conf.*, San Mateo CA, Morgan Kaufman.
106. Lucas E (1975) *Stochastic Convergence.* Academic Press, New York.
107. Mahfoud SW (1997) Niching Methods. Chapter C.6.1 in [15].
108. Manna Z (1974) *Mathematical Theory of Computation.* Mc Graw Hill, New York, St. Luis, San Francisco.
109. Martin WN, Lieing J, Cohon JP (1997) Island (Migration Models: Evolutionary Algorithms Based on Punctuated Equilibria). Chapter C.6.3 in [15].
110. Michalewicz Z (1996) *Genetic Algorithms + Data Structures = Evolutionary Programs.* Springer.
111. Michalewicz Z, Nazhiyath G, Michalewicz M (1966) A note of the usefulness of geometrical crossover for numerical optimization problems. In: eds. Fogel LJ, Angeline PJ, Bäck T *Proc. 5-th Ann. Conf. on Evolutionary Programming* MIT Press.
112. Miller BL (1997) *Noise, Sampling and Efficient Genetic Algorithms.* PhD Thesis, Univ. of Illinois at Urbana-Chamapaign.
113. Momot J, Kosacki K, Grochowski M, Uhruski P, Schaefer R (2004) Multi-Agent System for Irregular Parallel Genetic Computations. *LNCS* 3038, Springer, 623–630.
114. Neveu J (1975) *Discrete–Parameters Martingales.* North Holland, Amsterdam.
115. Nikolaev N, Hitoshi I (2000) Inductive Genetic Programming of Polynominal Learning Networks. In: Yao X ed. *Proc. of 1-th IEEE Symp. on Combinations of Evolutionary Computation and Neural Networks* ECNN-2000, IEEE Press, 158–167.
116. Nix E, Vose D (1992) Modeling Genetic Algorithms with Markov Chains. *Annals of Math. and Artificial Intelligence* 5(1):79–88.
117. Obuchowicz A (1997) The Evolutionary Search with Soft Selection and Deterioration of the Objective Function. In: *Proc. of 6-th Symp. Intelligent Information Systems and Artificial Intelligence IIS'97*, Zakopane, Poland, 288–295.
118. Obuchowicz A (1999) Adoption of the Time–Varying Landscape Using an Evolutionary Search with Soft Selection Algorithm. In: *Proc. of 3-rd KEGiOG Conf.*, Potok Złoty, Poland, 245–251.
119. Obuchowicz A (2003) Evolutionary Algorithms for Global Optimization and Dynamic System Diagnosis. University of Zielona Góra Press.
120. Obuchowicz A, Korbicz J (1999) Evolutionary Search with Soft Selection Algorithms in Parameter Optimization. In: *Proc. of the PPAM'99 Conf.*, Kazimierz Dolny, Poland, 578–586.
121. Obuchowicz A, Patan K (1997) An Algorithm of Evolutionary Search with Soft Selection for Training Multi-layered Feed Forwarded Networks. In: *Proc. of the*

3-rd Conf. Neural Networks and their Applications KSN'97, Kule, Poland, 123–128.
122. Obuchowicz A, Patan K (1997) About some Evolutionary Algorithm Cases. In: *Proc. of the 2-nd KAGiOG Conf.*, Rytro, Poland, 193–200 (in Polish).
123. Ombach J (1993) *Introduction to Probaility Theory.* Jagiellonian University Press, Textbooks 686 (in Polish).
124. Osyczka A (2002) *Evolutionary Algorithms for Single and Multicriteria Design Optimization.* Springer.
125. Pelczar A (1989) *Introduction to Theory of Differential Equations. Part II – Elements of the Quantitative Theory of Differential Equations.* PWN, Warsaw, Poland (in Polish).
126. Petty CC (1997) Diffusion (Cellural) Models. Chapter C.6.4 in [15].
127. Stopping Rules for the Multistart Method when Different Local Minima Have Different Function Values. *Optimization* 21.
128. Pintér JD (1996) *Global Optimization in Action.* Kluwer.
129. Podolak I, Telega H (2005) Hill crunching clustered genetic search and its improvements. *Nowy Sącz Academic Review* 2:9–69.
130. Podsiadło M (1996) Some Remarks About the Schemata Theorem. In: *Proc. of 1-th KAEiOG Conf.*, Murzasichle, Poland, 119–126 (in Polish).
131. Preprata FP, Shamos MI (1985) *Computational Geometry.* Springer.
132. Raptis S, Tzefastas S (1998) A Blueprint for a Genetic Meta-Algorithm. In: *Proc. 6-th European Conf. on Intelligent Techniques & Soft Computing*, Verlag Maintz, Vol. 1, 429–433.
133. Rechenberger I (1978) Evolutionsstrategien, Simulationsmethoden in der Medizin und Bioligie. In: Schneider B, Ranft U eds. *Simulationsmethoden in der Medizin und Biologie*, Springer, 83–114 (in German).
134. Reeves CR, Rowe JE (2003) *Genetic Algorithms: Principles and Perspectives. A Guide to the GA Theory.* Kluwer Academic Publishers.
135. Renders JM, Bersini H (1994) Hibridizing genetic algorithms with hill-climbing methods for global optimization: two possible ways. In: *Proc. of 1-st IEEE Conf. on Evolutionary Computation.* IEEE Press: 312–317.
136. Richardson M (1935) On the homology characters of symmetric products. *Duke Math. J.* 1 No. 1: 50–69.
137. Rinnoy Kan AHG, Timmer GT (1987) Stochastic Global Optimization Methods. Part 1: Clustering Methods. *Mathematical Programming* 39:27–56.
138. Rinnoy Kan AHG, Timmer GT (1987) Stochastic Global Optimization Methods. Part 2: Clustering Methods. *Mathematical Programming* 39:57–78.
139. Rosenberg RS (1967) *Simulation of Genetic Populations with Biomechanical Properties.* PhD Thesis, Univ. of Michigan, *Dissertation Abstracts International* 28(7), 2732B, (Univ. Microfilms No. 67–17, 836).
140. Rudolph G (1994) Convergence of Non-Elitist Strategies. In: *Proc. of 1-th IEEE Conf. on Computational Intelligence* Vol. 1, (Piscateway, NJ:IEEE), 63–66.
141. Rudolph G (1996) Convergence of Evolutionary Algorithms in General Search Spaces. In: *Proc. of 3-rd IEEE Conf. on Evolutionary Computations ICEC*, IEEE Press, 50–54.
142. Rudolph G (1994) How Mutation and Selection Solve Long Path Problem in Polynominal Expected Time. *Evolutionary Computation*, Vol. 2, No. 2, 207–211.
143. Rudolph G (1997) Stochastic Processes. Chapter B.2.2 in [15].

144. Rudolph G (1997) Models of Stochastic Convergence. Chapter B.2.3 in [15].
145. Rudolph G (2000) Evolution Strategies. Chapter 9 in [10].
146. Rzewuski M, Szreter M, Arabas J (1997) Looking for the More Effective Operators for Evolution Strategies. In: *Proc. of the 2-nd KAGiOG Conf.*, Rytro, Poland, 237–243 (in Polish).
147. Schaefer R (2000) Adaptability and Self-Adaptability in Genetic Global Optimization. In: *Proc. of AIMETH'00*, Gliwice, Poland, 291–298.
148. Schaefer R (2001) Simple Taxonomy of the Genetic Global Optimization. *Computer Assisted Mechanics and Engineering Sciences CAMES* 9:139–145.
149. Schaefer R (with the chapter 6 written by Telega H) (2002) An Introduction to the Global Genetic Optimization. Jagiellonian University Press (jn Polish).
150. Schaefer R (2003) Essential Features of Genetic Strategies. In: *Proc. of the CMM'03 Conf.*, Wisła, Poland, 41–42.
151. Schaefer R (2004) Detailed Evaluation of the Schemata Cardinality Modification at the Single Evolution Step. In: *Proc. of the 7-th KAEiOG Conf.*, Kazimierz, Poland, 143–147.
152. Schaefer R, Adamska K, Telega H (2004) Genetic Clustering in Continuous Landscape Exploration. *Engineering Applications of Artificial Intelligence* EAAI, Elsevier, 17:407–416.
153. Schaefer R, Adamska K (2004) Well-Tuned Genetic Algorithm and its Advantage in Detecting Basins of Attraction. In: *Proc. of the 7-th KAEiOG Conf.*, Kazimierz, Poland, 149–154.
154. Schaefer R, Jabłoński ZJ (2002) On the Convergence of Sampling Measures in Global Genetic Search. *LNCS* 2328, Springer, 593–600.
155. Schaefer R, Jabłoński ZJ (2002) How to Gain More Information from the Evolving Population? Chapter in: Arabas J ed. *Evolutionary Computation and Global Optimization*, Warsaw Technical University Press, 21–33.
156. Schaefer R, Kołodziej J, Gwizdała R, Wojtusiak J (2000) How Simpletons can Increase the Community Development – an Attempt to Hierarchical Genetic Computation. In: *Proc. of 4-th KAEiOG Conf.*, Lądek Zdrój, 187–198.
157. Schaefer R, Kołodziej J (2003) Genetic Search Reinforced by the Population Hierarchy. In: De Jong KA, Poli R, Rowe JE eds. *Foundations of Genetic Algorithms 7*, Morgan Kaufman, 383–399.
158. Schaefer R, Telega H, Kołodziej J (1999) Genetic Algorithm as a Markov Dynamic System. In: *Proc. of the Int. Conf. on Intelligent Techniques in Robotics, Control and Design Making*, Polish-Japanese Institute of Information Technology Press, Warsaw.
159. Schaffer JD, Morishima A (1987) An Adoptive Crossover Distribution Mechanism for Genetic Algorithm. In: Greffenstette JJ, Hillsdale NJ eds. *Proc. of 2-nd Int. Conf. on Genetic Algorithms*, Erlbaum, 36–40.
160. Schlierkamp-Voosen D, Müchlenbein H (1996) Adaptation of Population Sizes by Competing Subpopulation. *Proc. of 1-th IEEE Conf. on Evolutionary Computation*, 330–335.
161. Schraudolph NN, Belew RK (1992) Dynamic parameter encoding for genetic algorithms. *Machine Learning Journal*, Volume 9, Number 1, 9–21.
162. Schwartz L (1967) *Analyse Mathematique.* Hermann, Paris (in French).
163. Schwefel HP (1977) Numerische Optiemierung von Computer Modellen Mittels der Evolutionsstrategie. *Interdisciplinary System Research* 26, Birkhäuser, Basel (in German).

164. Schwefel HP, Bäck T (1977) Artificial Evolution: How and Why? In: *Proc. of EUROGEN'97*, Willey, 1–19.
165. Schwefel HP, Rudolph G (1995) Contemporary Evolution Strategies. Advances in Artificial Life. In: Morgan F ed. *Proc. of 3-rd Conf. in Artificial Life, LNCS* 928, 893–907.
166. Seredyński F (1998) New Trends in Parallel and Distributed Evolutionary Computing. *Fundamenta Informaticae* 35:211–230.
167. Shaw JEH (1988) A Quasirandom Approach to Integration in Bayesian Statistics. *The Annals of Statistics* 16:895–914.
168. Skiena S (2003) *Implementing Discrete Mathematics: Combinatorics and Graph Theory with Mathematica.* Cambridge University Press.
169. Skolicki Z, De Jong K (2004) Improving Evolutionary Algorithms with Multi-representation Island Models, *Proc. of 8-th Int. Conf. on Parallel Problem Solving from Nature - PPSN VIII*, Birmingham, UK, *LNCS*, Vol. 3242, Springer, 420–429.
170. Skolicki Z, De Jong K (2005) The influence of migration sizes and intervals on island models, *Proc. of Genetic and Evolutionary Computation Conference - GECCO-2005*, Washington DC, ACM Press, 1295–1302.
171. Slavov V, Nikolaev NI (1997) Inductive Genetic Programming and the Superposition of Fitness Landscapes. In: Bäck T ed. *Proc. of 7-th Int. Conf. on Genetic Algorithms*, ICGA-97, Morgan Kaufman, 97–104.
172. Slavov V, Nikolaev NI (1999) Genetic Algorithms, Fitness Sublandscapes and Subpopulations. In: Banzhaf W, Reeves C eds. *Foundations of Genetic Algorithms* 5, Morgan Kaufman.
173. Smith PA (1933) The topology of involutions. In: *Proc. Nat. Acad. Sci.* 19, No. 6: 612–618.
174. Smith S (1998) The Simplex Method and Evolutionary Algorithms. In: *Proc. of ICEC'98 Conf.*, Alaska, USA, IEEE Press.
175. Sobol IM (1982) On the Estimate of the Accuracy of a Simple Multidimensional Search. *Soviet. Math. Dokl.* 26:398–401.
176. Sobol IM, Statnikow RB (1981) *The Selection of Optimal Paraleters in the Multiparameter Problems.* Nauka, Moskva (in Russian).
177. Spears WM (1994) Simple Subpopulation Systems. In: *Proc. of 3-rd Annual Conf. on Evolutionary Programming*, World Scientific, 296–307.
178. Spears WM (2000) *Evolutionary Algorithms.* Springer.
179. Spivak M (1969) *Calculus on Manifolds.* W.A. Benjamin, Inc. New York, Amsterdam.
180. Stadler P 1995 Towards a theory of Landscapes. In: Lopéz-Pena, Capovilla R, Garcia-Pelayo R, Waelbrock H, Zertouche F eds. *Complex systems and Binary Networks*, Springer, 77–173.
181. Stańczak J (1999) *The Development of the Algorithm Concept for Self-Refinement of Evolutionary Systems.* PhD Thesis, Warsaw Technical University, Warsaw, Poland.
182. Stańczak J (2000) Evolutionary Algorithms with the Population of "Intelligent" Individuals. In: *Proc. of 4-th KAEiOG Conf.*, Lądek Zdrój, 207–218 (in Polish).
183. Sukharev AG (1971) Optimal Strategies of the Search for the Extremum. *Computational Mathematics and Mathematical Physics* 11:119–137.
184. Telega H (1999) *Parallel Algorithms for Solving Selected Inverse Problems.* PhD Thesis, AGH University of Science and Technology, Kraków, Poland.

185. Telega H, Schaefer R (1999) Advances and Drawbacks of a Genetic Clustering Strategy. In: *Proc. of 3-rd KEGiOG Conf.*, Potok Złoty, Poland, 291–300.
186. Telega H, Schaefer R (2000) Testing the Genetic Clustering with SGA Evolutionary Engine. In: *Proc. of 4-th KAEiOG Conf.*, Lądek Zdrój, Poland, 227–263.
187. Thierens D (1995) *Analysis and Design of Genetic Algorithms.* PhD Thesis, Katholieke Univ. Leuven, Leuven, Belgium.
188. Törn A (1975) A Clustering Approach to Global Optimization. In [60].
189. Törn A (1976) Cluster Analysis Using Seed Points and Density Determined Hyperspheres with an Application to Global Optimization. In: *Proc. of the 3-rd Int. Conf. of Pattern Recognition*, Colorado, California, 394–398.
190. Törn A, Viitanen S (1994) Topographical Global Optimization Using Pre-Sampled Points. *Journal of Global Optimization* 5:267–276.
191. Vose MD (1996) Modeling Simple Genetic Algorithms. *Evolutionary Computation* 3(3):453–472.
192. Vose MD (1997) Logarithmic Convergence of Random Heuristic Search. *Evolutionary Computation* 4(4):395–404.
193. Vose MD (1999) *The Simple Genetic Algorithm*,MIT Press.
194. Vose MD, Liepnis GE (1991) Punctuated Equilibria in Genetic Search. *Complex Systems* 5:31–34.
195. Vose MD, Wright AD (1995) Stability of Vertex Fixed Points and Applications. *Evolutionary Computation.*
196. Whitley D (1994) A Genetic Algorithm Tutorial. *Statistics and Computing* 4:65–85.
197. Whitley D, Gordon VS, Mathias K (1994) Lamarckian Evolution, the Baldwin Effect and Function Optimization. In: Davidor Y, Schwefel HP, Mäner R eds. *Lecture Notes in Computer Science* 866:6–15 Berlin, Springer.
198. Whitley D, Gordon VS (1993) Serial and Parallel Genetic Algorithms as Function Optimizers. In: Forrest S ed. *Proc. of ICGA'97*, Morgan Kaufman, San Mateo, CA, 177–183.
199. Whitley D,Mathias K, Fitzhorn P (1991) Delta Coding: An Iterative search Strategy for Genetic Algorithms. In: Belew RK, Booker LB eds. *Proc. of the 4-th Int. Conf. on Genetic Algorithms.* Morgan Kaufman, San Mateo, CA, 77–84.
200. Whitley D, Soraya R, Heckerdorn RB (1997) Island Model Genetic Algorithms and Linearly Separable Problems. In: *Proc. of AISB'97 Workshop on Evolutionary Computing*, Manchester, 112–129.
201. Wierzba B, Semczuk A, Kołodziej J, Schaefer R (2003) Hierarchical Genetic Strategy with real number encoding. In: *Proc. of 6-th KAEiOG Conf.* Łagów Lubuski, Poland, 231–237.
202. Wit R (1986) *Nonlinear Programming Methods.* WNT, Warsaw (in Polish).
203. Wright AH (1994) Genetic Algorithms for Real Parameter Optimization. In: Davis L ed. *Foundations of Genetic Algorithms*, Morgan Kaufman: 205–218.
204. Wright AH (2000) The Exact Schema Theorem. Technical report, University of Montana, Missoula, MT 59812, USA, 1999. http://www.cs.umt.edu/u/wright/. http://citeseer.ist.psu.edu/article/wright99exact.html
205. Wright AH, Rowe J, Stephens Ch, Poli R (2003) Bistability in a GenePool GA with Mutation. In: De Jong KA, Poli R, Rowe JE eds. *Foundations of Genetic Algorithms 7*, Morgan Kaufman, 63–80.

206. Yen J, Liao J, Lee B, Rendolph D (1998) A Hybrid Approach to Modeling Metabolic Systems Using a Genetic Algorithm and Simplex Method. *IEEE Transactions on Systems. Man and Cybernetics. Part B: Cybernetics* 28:173–191.
207. Zeidler E (1985) *Nonlinear Functional Analysis and its Application.* (II.1.A. *Linear Monotone Operators*, II.1.B. *Nonlinear Monotone Operators*, III. *Variational Methods and Optimization*).
208. Zieliński R (1981) A Stochastic Estimate of the Structure of Multi-Extremal Problems. *Math. Programming* 21:348–356.
209. Zieliński R, Neuman P (1986) *Stochastic Methods of Searching Function Minima.* WNT, Warsaw (in Polish).
210. Zienkiewicz OC, Taylor R (1991) *The Finite Element Method* Vol. 1, 2 (fourth edition), McGraw-Hill, London.

Index

Printing: Krips bv, Meppel
Binding: Stürtz, Würzburg